A

THEORETICAL AND PRACTICAL

ARITHMETIC;

DESIGNED FOR

COMMON SCHOOLS AND ACADEMIES.

BY

DANIEL LEACH AND WILLIAM D. SWAN.

PHILADELPHIA:

THOMAS, COWPERTHWAIT, & CO.

1851.

STEREOTYPED AT THE
BOSTON STEREOTYPE FOUNDRY.

PREFACE.

It has been the aim of the authors, in preparing this work, to make it eminently both a practical and a theoretical treatise on the science of numbers. They have therefore adopted that arrangement which has appeared the most philosophical, and, at the same time, the best suited to the comprehension of the learner.

They have bestowed great labor on the rules and definitions, in order to make them *lucid, concise,* and *accurate;* and they have carefully avoided introducing any illustrations or remarks not necessary to a clear understanding of the subject.

The examples have been prepared with much discrimination. Many of them are questions which have actually occurred in ordinary business transactions.

They would call attention first to the examples in addition, some of which have been so arranged as to bring together the same combination of figures throughout the same line. The long leger columns are designed for those who wish to acquire a facility in adding long columns. This is one of the most useful exercises in arithmetic to which pupils can be accustomed.

They would also call particular attention to the rule for finding the least common multiple, the rule of alligation, and the rule for extracting the cube root. These rules are clear and concise, and can be most rigidly demonstrated; while in the processes indicated by them there is a saving of more than one half of the figures, as compared with the processes in similar works now in common use. The section on fractions, they think, will also commend itself to every experienced teacher.

Although the preface is not the proper place for discussing the best method of teaching arithmetic, yet the authors cannot refrain from urging upon all teachers not to allow their pupils to attempt to solve a question till they fully understand all its conditions, and *always* to require them to state the principles upon which each solution is founded. Pupils should be accustomed to write questions of their own under each rule. This is a very important exercise.

They would also suggest that, in *every* question in which there are both multiplication and division, the pupil should at first indicate the processes by their appropriate signs, and then cancel the factors common to the dividend and divisor.

In the preparation of this work, a large number of the recent English and American treatises on arithmetic have been consulted, and the most valuable mathematical works of the French.

CONTENTS.

A

THEORETICAL AND PRACTICAL

ARITHMETIC.

INTRODUCTION.

SECTION I.

ART.

1. ARITHMETIC is the science of numbers.

2. *Numbers* express how many units, or parts of a unit, there are in any quantity.

3. *Quantity* is any thing that can be increased, diminished, or measured.

4. The least whole number employed to express or measure quantity of the same kind is called a *unit.*

5. A number expressing a particular kind of a unit is called a *concrete* number; as, 1 dollar, 2 books, &c.

6. When a number does not express any particular kind of a unit, it is called an *abstract* number; as, 1, 4, 7, &c.

OBS. 1. Concrete numbers are also called *denominate* or *compound* numbers.

OBS. 2. Whole numbers are sometimes called *integers.*

7. In the computation of numbers, ten characters are employed, called *figures;* thus: 1, one; 2, two; 3,

---

What is arithmetic? What are numbers? What is quantity? What is a unit? What is a concrete number? What is an abstract number?

three; 4, four; 5, five; 6, six; 7, seven; 8, eight; 9, nine; 0, cipher. The first nine figures are called *significant*, because they have a given value assigned them. The cipher has no representative value, but is used where no number is to be expressed.

OBS. 1. By the aid of these ten characters any possible or conceivable quantity may be expressed.

OBS. 2. The first nine figures are sometimes called *digits*, from the Latin word *digitus*, signifying a *finger*.

8. The various operations of arithmetic are performed by NUMERATION, ADDITION, MULTIPLICATION, SUBTRACTION, and DIVISION.

Addition and multiplication are employed to show how numbers may be *increased*; subtraction and division how they may be *diminished*.

NUMERATION.

SECTION II.

9. NUMERATION is the art of expressing any number whatever by figures.

Figures are arranged in different orders or places, and have different values assigned them, according to the place they occupy. The first place, which is always at the right, represents units; the second, tens; the third, hundreds; the fourth, thousands; the fifth, tens of thousands; the sixth, hundreds of thousands; the seventh, millions, &c. Thus the figure 1 represents a unit, a ten, a hundred, a thousand, &c., according to the place it occupies. In all places in which no number is to be expressed, ciphers must be written.

Which are significant figures? How are the various operations of arithmetic performed? What is numeration? How are figures arranged? What does the first place represent? The second? The third? The fourth? The fifth? &c. When must ciphers be used?

Thus, if any figure be written in the fourth place, or the place of thousands, a cipher must be written in th place of units, and in the place of tens, and in place of hundreds; as, 7000.

It is evident that the same figure represents in the second place a value ten times greater than in the first place, and in the third place ten times greater than in the second, and a hundred times greater than in the first; and by each removal of any figure to the next place on the left its representative value is increased ten times.

10. The different places in which figures are arranged may be divided into periods of three figures each.

NUMERATION TABLE.

Place	Figure	Period
Hundreds of quintillions.	6	7th period.
Tens of quintillions.	4	
Quintillions.	9	
Hundreds of quadrillions.	8	6th period.
Tens of quadrillions.	7	
Quadrillions.	6	
Hundreds of trillions.	5	5th period.
Tens of trillions.	4	
Trillions.	8	
Hundreds of billions.	7	4th period.
Tens of billions.	1	
Billions.	9	
Hundreds of millions.	8	3d period.
Tens of millions.	7	
Millions.	6	
Hundreds of thousands.	5	2d period.
Tens of thousands.	4	
Thousands.	3	
Hundreds.	6	1st period.
Tens.	0	
Units.	1	

The number indicated in the above table is read thus: six hundred and forty-nine *quintillions;* eight

What effect is produced on the representative value of a figure by changing its place?

hundred and seventy-six *quadrillions ;* five hundred and forty-eight *trillions ;* seven hundred and nineteen *billions ;* eight hundred and seventy-six *millions ;* five hundred and forty-three *thousand ;* six hundred and one.

11. Read the following numbers: —

1.	12	7.	1010101010
2.	21	8.	2020202020
3.	123	9.	6006006006
4.	1010	10.	2222222222
5.	10209	11.	54467841234
6.	202807	12.	123456789000

12. Express in figures the following numbers: —

1. Twenty-four.
2. Two hundred and four.
3. Two hundred and forty.
4. Two thousand and four.
5. Two thousand and forty.
6. Forty-six thousand, five hundred and twenty.
7. Four hundred and six thousand, five hundred and two.
8. Eight hundred thousand, one hundred and one.
9. One million, one thousand, one hundred and one.
10. Ten millions, ten thousand, and ten.
11. One hundred millions, one hundred thousand, and one hundred.
12. Two millions, six hundred and ten thousand, four hundred and forty-six.
13. Sixty-four millions, nine hundred and ten.
14. Two hundred and forty millions, three thousand.
15. Five hundred and sixty-seven billions, three hundred and forty-eight millions, seven hundred and twenty thousand, six hundred and forty.
16. Fourteen trillions, six billions, three hundred and forty millions, and twenty-two.

ADDITION.

SECTION III.

13. ADDITION is the process of finding the sum of two or more numbers of the same kind.

OBS. The sum expresses the total value of the several numbers, or as many units as there are in all of them.

14. RULE. *Write the numbers under each other, units under units, tens under tens, hundreds under hundreds, &c. First, add the column of units, and write under this column the right hand figure of the sum, and add the remaining figure or figures to the next column. Add all the columns in the same manner, and under the last write the whole sum contained in it.*

PROOF. *Beginning at the top of the column of units, add each column downwards; and if the result be the same as the first, the work is supposed to be right.*

15. Two signs are often employed in addition; the one, $+$, called *plus*, which signifies *added to*, or *and*, and the other, $=$, called the *sign of equality*, which signifies *equal to*, or *are:* thus, $4 + 6 = 10$ is read, four and six are ten.

1. Add together the following numbers:—

```
  2472
 51856
 27692
 70347
 81850
------
234217
```

The sum of the first column is 17; the 7 is written

What is addition? What does the sum express? What signs are used in addition? Recite the rule.

under the column of units, and the 1 is added to the next column, whose sum is 31; the 1 is written under the column of tens, and the 3 is added to the next column, whose sum is 32; the 2 is written under the column of hundreds, and the 3 is added to the next column, whose sum is 14; the 4 is written under the column of thousands, and the 1 is added to the next column, whose sum is 23, which is written under the last column.

16. It is evident that adding the left hand figure to the next column is the same as adding the tens in the column of units, the hundreds in the column of tens, the thousands in the column of hundreds, &c., which may be illustrated as follows: —

2.

```
  465
  847
  963
 ----
   15 units.
  16  tens.
 21   hundreds.
 ---------------
 2275 sum.
```

The sum of the first, or the column of units, is 15, which is 1 ten and 5 units; the 5 is written under the column of units, the 1 under the tens. The sum of the second, or the column of tens, is 16, which is 1 hundred and 6 tens; the 6 is written under the column of tens, and the 1 under the column of hundreds. The sum of the third, or the column of hundreds, is 21, which is 2 thousand and 1 hundred; the 1 is written under the column of hundreds, and the 2 is written in the next place on the left, which is the place of thousands. These added together give the sum total 2275.

What is the effect of adding the left hand figure to the next column?

17. This rule depends upon the principle illustrated in Article 9th, that ten in the column of units is equal to one in the column of tens, and ten in the column of tens is equal to one in the column of hundreds, &c.

EXAMPLES.

3.	4.	5.	6.	7.
2222	2222	2222	2222	2222
3333	3333	3333	3333	3333
4445	5556	6667	7778	8889

8.	9.	10.	11.	12.
3333	4444	5555	6666	7777
4444	5555	4444	3333	4444
7778	6667	7778	8889	7779

13.	14.	15.	16.	17.
3333	6666	5555	8888	3333
4444	5555	4444	2222	9999
5555	4444	7777	6666	4444
6666	3333	4444	3333	7777
1113	2224	1113	3335	0002

18.	19.	20.	21.	22.
3333	4444	5555	6666	9999
8765	8765	9876	9876	8765
3333	4444	5555	6666	9999
8765	8765	9876	9876	8765
3333	4444	5555	6666	9999
6545	5445	6546	5546	4315

Upon what principle does the rule of addition depend?

23. Add together 12345, 54321, 678, and 876.
24. Add together 6789, 3211, 7456, and 2546.
25. Add together 40396, 43745, 675, and 96.
26. Add together 70964, 84345, 4327, and 75.
27. Add together 90, 4360, 10345, and 467634.
28. Add together 3, 7506, 42, 90704, and 736.
29. Add together 78960, 45, 960, 301, and 84.
30. Add together 5360, 4263, 2, 7503, and 801.
31. Add together 934, 8375, 2013, 46, and 7640.
32. Add together 7560, 473, 421, 76, and 96341.
33. Add together 7345, 8403, 642, 24, and 8460.
34. Add together 9345, 6704, 6340, 54, and 760.
35. Add together 2317, 5742, 96043, and 8483.
36. Add together 2319, 64893, 45407, and 98.
37. Add together 4296, 3715, 39624, and 7434.
38. Add together 968732, 201, 28346, and 291.
39. Add together 234604, 8764, 2346, and 734.
40. Add together 424603, 26934, 80973, and 73.
41. 964564+843453+372131+234560=?
42. 843735+345841+673450+343466=?
43. 432846+648345+873459+673489=?
44. 849321+734463+734649+763921=?
45. 734561+234764+245321+289641=?
46. 346734+824375+654342+734931=?
47. 729329+324643+320103+345603=?
48. 678456+324321+735920+275463=?
49. 372635+424851+734639+273486=?
50. 934275+604922+234769+832834=?
51. 634296+234283+864234+234675=?
52. 376263+456934+963264+239340=?
53. 763412+123456+789654+236960=?
54. 656789+756789+639936+735535=?
55. 324423+459954+863638+450207=?
56. 875634+845673+437734+960784=?
57. 735640+937456+784503+875406=?
58. 345634+783453+456278+673840=?

59. What is the sum of the following numbers? — Six hundred and five; thirty-seven; four thousand five hundred and twenty; thirty-seven millions; two hundred and one; ninety-nine thousand and nine.

60. Find the sum of three thousand seven hundred and forty-four; nine million fourteen thousand and eleven; five hundred and eight millions two hundred and three thousand and twenty-five.

61. A merchant, commencing business, had in cash 6330 dollars, goods valued at 9875 dollars, bank stock valued at 4320 dollars, railroad stock valued at 2700 dollars: during the year, he gained above his expenses 2316 dollars. What was he worth at the end of the year?

62. A merchant sold five bales of cloth. For the first bale he received 735 dollars, for the second 637 dollars, for the third 573 dollars, for the fourth 391 dollars, for the fifth 721 dollars. How much did he receive?

63. A farmer received for the produce of his farm in one year as follows: for hay 276 dollars, for potatoes 391 dollars, for oats and corn 234 dollars, for fruit 567 dollars. How much did he receive?

64. A man paid 3234 dollars for his farm, 5640 dollars for his house, 1500 dollars for his furniture, and 539 dollars for his stock and tools. What did he pay for the whole?

65. There are two numbers; the less is 93078, the difference is 4796. What is the greater?

66. A man owns three farms; the first is valued at 5697 dollars, the second is valued at 9630 dollars, the third at 1639 dollars. How much are the three worth?

67. A gentleman left in his will to his three sons 1930 dollars each, to his two daughters 1737 dollars each, to his wife 930 dollars more than all his children. What was his wife's portion, and what was the value of the whole estate?

68.	69.	70.	71.
76543	98765	76586	92345
76543	98765	34524	18765
76543	98765	76586	92345
76543	98765	34524	18765
76543	98765	76586	92345
76543	98765	34524	18765
76543	98765	76586	92345
76543	98765	34524	18765
76543	98765	76586	92345
76543	98765	34524	18765
76543	98765	76586	92345
76543	98765	34524	18765
76543	98765	76586	92345
76543	98765	34524	18765
54943	65555	47858	76425

72.	73.	74.	75.
33146	29956	45353	32142
40354	97194	88868	88459
37644	11613	22242	26976
72212	82453	84587	32785
31254	27644	86523	72329
27623	93216	48868	86788
53433	19259	82242	21863
37654	81621	24587	33369
12315	27434	83523	76785
61141	91956	13868	84956
37654	12159	14242	26479
31234	97695	84587	31787
42222	23416	22523	73324
37654	87999	88867	86789
44454	21984	16818	67809

76.	77.	78.	79.
244658	275634	135790	123456
492327	386731	246824	789123
635425	987654	135790	456789
321465	321456	864212	123456
732849	989123	579246	788123
376731	456789	835792	456789
935746	123456	468357	123456
847963	789123	924689	789123
745143	456789	753246	456789
234561	123456	835792	123456
746874	789123	468357	789123
934746	456789	924683	456789
872345	123456	579246	123456
934756	789123	835798	789123
842345	456789	642875	456789
873456	123456	324683	123456
864580	789123	579864	789122
234672	456789	297531	456789
325871	246842	135795	871178
479234	357931	246834	936639
845645	642248	824248	248842
823456	756139	357964	525255
245734	246842	872278	736376
872475	657931	375946	875578
896731	642248	624862	473468
456841	753139	375937	934579
314567	246842	872459	894645
814563	357931	837645	123875
427831	642248	644875	767457
932768	753913	472963	875345
456345	375913	875847	874563
345634	426428	864314	375534
734734	573931	734561	937565
734564	624824	273475	875734
834756	735913	845675	698945

MULTIPLICATION.

SECTION IV.

18. Multiplication is the process of finding the sum of any number, when taken as many times as there are units in another number.

Obs. This definition of multiplication is applicable only to whole numbers.

The number produced by multiplication is called the *product.* The number multiplied is called the *multiplicand.* The number multiplied by is called the *multiplier.* The multiplicand and multiplier are called *factors.*

Obs. The term *factor* is derived from a Latin word, signifying to *make*, or *produce.*

19. The sign of multiplication is a cross, ×, and is read, multiplied by, or times. Thus, 9×8=72 is read, 9 times 8 are 72. 9 is the multiplicand, 8 the multiplier, and 72 the product.

20. Rule. *Write the multiplier under the multiplicand, units under units, tens under tens, &c. When there is but one figure in the multiplier, multiply each figure in the multiplicand, beginning with units, and write the right hand figure of the product under its multiplier, and add the remaining figure or figures as in addition. When there is more than one figure in the multiplier, multiply each figure in the multiplicand, beginning with units, by each figure of the multiplier in succession, and write the right hand figure of the product under its multiplier, and add the left hand figure or figures as in*

What is multiplication? What is the number produced by multiplication called? What is the multiplicand? What is the multiplier? What are factors? What is the sign of multiplication? Recite the rule.

addition. The sum of the several products will be the product sought.

PROOF. *Make the multiplicand the multiplier, and the multiplier the multiplicand, and, if their product be the same as before, the work is supposed to be right.*

It is the most convenient to take the larger number for the multiplicand, and the smaller for the multiplier. The result will be the same, whether the larger number be repeated as many times as there are units in the smaller, or the smaller as many times as there are units in the larger. Either the multiplicand or multiplier must always be considered as an abstract number, for it is absurd to suppose that one number can be repeated as many times as there are pounds, yards, or dollars.

1. Multiply 4657 by 8.

```
 4657
    8
-----
37256
```

The 7 units, multiplied by 8, are equal to 56: write the 6 under the 8, and reserve the 5 to be added to the next product. 8 times 5 are 40, to which 5 being added, makes 45: write the 5 under the column of tens, and reserve the 4 to be added to the next product. 8 times 6 are 48, to which 4 being added, makes 52: write the 2 under the column of hundreds, and reserve the 5 to be added to the next product. 8 times 4 are 32, to which 5 being added, makes 37, which is to be written under the last column.

2. Multiply 65943 by 58.

```
  65943
     58
-------
 527544
329715
-------
3824694
```

Multiply by the 8 units, as in the last example. Then multiply by the 5 tens. The first product is 15: write the 5 directly under its multiplier, reserve the 1 for the next product of 4 by 5, which is 20; the 1 being added, makes 21: write the 1 at the left of the 5, under the column of hundreds, and reserve the 2 for the next product of 9 by 5, which is 45, to which 2 being added, makes 47: write the 7 at the left of the 1, and reserve the 4 for the next product of 5 by 5, which is 25, to which 4 being added, makes 29: write the 9 at the left of the 7, and reserve the 2 for the next product of 6 by 5, which is 30, to which 2 being added, makes 32, which is to be written at the left of the 9. The sum of the several products is 3824694, the required product.

EXAMPLES.

3.	4.	5.	6.
4294	3276	8752	9743
5	4	6	9

7.	8.	9.	10.
8349	2546	3767	5886
24	32	44	58

11.	12.	13.	14.
9654	4569	8756	6643
364	263	942	842

15.	16.	17.	18.
2756	3256	9463	3472
436	384	272	357

19. 934567 × 13 = ?	29. 456789 × 4361 = ?
20. 845603 × 14 = ?	30. 865432 × 5624 = ?
21. 945679 × 15 = ?	31. 936745 × 6345 = ?
22. 843475 × 16 = ?	32. 845346 × 7346 = ?
23. 975349 × 17 = ?	33. 934575 × 8431 = ?
24. 873945 × 18 = ?	34. 654321 × 3214 = ?
25. 943457 × 19 = ?	35. 897654 × 4576 = ?
26. 896321 × 21 = ?	36. 356891 × 5316 = ?
27. 456789 × 23 = ?	37. 896453 × 6452 = ?
28. 876432 × 24 = ?	38. 943456 × 9345 = ?

21. When there are ciphers at the right hand of either the multiplicand or multiplier, or both, —

RULE. *Write the significant figures under each other, and multiply only by the significant figures, and to the product annex as many ciphers as there are at the right hand of both of the factors.*

39. 6456×20 = ?	42. 45634×6200 = ?
40. 3400×7300 = ?	43. 75000×32000 = ?
41. 46000×9300 = ?	44. 8400×7300 = ?

OBS. Multiplying by 10 is the same as annexing one cipher to the multiplicand; multiplying by 100, the same as annexing two ciphers; by 1000, the same as annexing three ciphers, &c.

22. When there are ciphers between the significant figures of the multiplier, —

RULE. *Multiply by the significant figures only, and write the first product of each figure directly under its multiplier.*

45. 7644×304 = ?	47. 96964×3004 = ?
46. 8456×40006 = ?	48. 97564×20008 = ?

What is the rule for multiplication when there are ciphers at the right of the multiplicand, multiplier, or both? What is the rule when there are ciphers in the multiplier?

23. When the multiplier is a composite number, —

RULE. *Separate the multiplier into its several factors, and multiply first by one factor, and that product by another, and so on till all the factors have been used as multipliers. The last product will be the answer.*

OBS. A composite number is one which may be produced by multiplying together two or more numbers, both of which must be greater than a unit. The numbers producing the composite number, when multiplied together, are called *factors*. (ART. 17.) Thus, 36 is a composite number, and may be produced by multiplying together 3 and 12, 6 and 6, 4 and 9, 18 and 2, and 3 and 3 and 4.

49. Multiply 6789 by 42 or 7×6 or $3\times2\times7$.
50. Multiply 7845 by 56 or 8×7 or $4\times2\times7$.
51. Multiply 8643 by 54 or 9×6 or $3\times3\times3\times2$.
52. Multiply 9368 by 81 or 9×9 or $3\times9\times3$.
53. Multiply 8645 by 64 or 8×8 or $2\times4\times8$.

24. When the multiplier consists wholly of 9's, —

RULE. *Annex to the multiplicand as many ciphers as there are 9's in the multiplier, and from this number subtract the multiplicand.*

25. This rule depends upon the principle, that annexing one cipher to the multiplicand multiplies it by 10; that annexing two ciphers, multiplies it by 100, &c. Now, it is evident that if the multiplicand be multiplied by 10, it is repeated once too many times for the product of the multiplicand by 9; if multiplied by 100, it is repeated once too many times for the product of the multiplicand by 99, &c. (ART. 28.)

54. Multiply 67899 by 9999.

```
678990000
    67899
---------
678822101
```

What is the rule when the multiplier is a composite number? What is a composite number? What is the rule when the multiplier consists wholly of 9's?

55. What will 565 barrels of flour cost, at 7 dollars a barrel?

56. How many bushels of corn will 96 acres of land produce, if each acre produces 33 bushels?

57. In one bushel there are 32 quarts. How many quarts are there in 156 bushels?

58. In one year there are 8766 hours. How many hours will a boy have lived when he is 12 years old?

59. In one year there are 525960 minutes. How many minutes will a boy have lived when he is 15 years old?

60. In one acre of land there are 43560 feet. What would be the price of 1 acre, at 7 cents per foot? What at 8 cents? What at 12 cents? What at 25 cents?

61. Two men start from the same place, at the same time, and travel the same way; the one travels 36 miles a day, and the other 45 miles a day. How far apart will they be in 9 days?

62. Two men start from the same place, at the same time, and travel in opposite directions; the one travels 9 hours a day, at 7 miles an hour; the other travels 12 hours a day, at 8 miles an hour. How far apart will they be in 6 days? How far in 11 days?

63. If a railroad car moves 38 miles an hour, how far would it go in 30 days, of 24 hours each, allowing 2 hours each day for stopping?

64. If 9 men can do a piece of work in 13 days, how long would it take one man to do the same work? How many men would do it in one day?

65. In 1 mile there are 320 rods. How many rods are there in 42 miles? In 336 miles?

66. A merchant bought 54 pieces of cloth, each piece containing 39 yards, and paid 5 dollars a yard? What did he give for the whole?

67. In one day there are 1440 minutes. How many minutes in 365 days?

SUBTRACTION.

SECTION V.

26. Subtraction is the process of finding the difference between two numbers of the same kind.

The larger number is called the *minuend;* the smaller, the *subtrahend.*

Obs. The term *minuend* is from a Latin word signifying to be *diminished; subtrahend* from a Latin word signifying to be *taken from.*

27. The sign employed in subtraction is a short horizontal line, — and is read *less.* Thus 6 — 4 = 2 is read, 6 less 4 is equal to 2.

28. Rule. *Write the less number under the greater, units under units, tens under tens, &c. Beginning with units, subtract each figure in the lower line from the one above it, and write underneath their difference. If the figure in the lower line be greater than the one above it, add* 10 *to the upper figure before subtracting, and add* 1 *to the next left hand figure in the lower line.*

Proof. *Add the remainder to the smaller number, and, if the work be right, it will be equal to the larger.*

29. This rule depends upon the evident principle, that the difference of two numbers remains the same when each of them is increased by the addition of any given number; and the adding of 10 to any column, in the upper line, is the same as adding 1 to the next left hand column, in the lower line, (Art. 9.)

1. From 8434536 take 4530644.

```
8434536
4530644
-------
3903892 = the difference.
```

What is subtraction? What is the minuend? What the subtrahend? From what are they derived? What is the sign of subtraction? Recite the rule.

4 units from 6 units leaves 2 units, which is written underneath. As 4 cannot be taken from 3, 10 is added to the 3, making 13. 4 from 13 leaves 9, which is written underneath, and 1 is added to the 6, the next figure on the left, in the lower line, making 7. As 7 cannot be taken from 5, 10 is added to the 5, making 15. 7 from 15 leaves 8, which is written underneath, and 1 is added to the cipher in the lower line, making 1, which taken from 4 leaves 3. 3 taken from 3 leaves 0, which is written underneath. As 5 cannot be taken from 4, 10 is added, which makes 14. 5 from 14 leaves 9, which is written underneath. 1 added to the 4 makes 5; 5 from 8 leaves 3.

EXAMPLES.

2.	3.	4.
987654	404045	678932
123456	204024	456734
5.	**6.**	**7.**
456789	573456	875678
357901	345674	734567
8.	**9.**	**10.**
100000	1010101	9090909
1	10	90
11.	**12.**	**13.**
100000	100000	100000
33334	44445	55556
14.	**15.**	**16.**
4040404	6060606	8080808
404040	606060	808080

17. 86401 — 7356 = ?	23. 960304 — 730245 = ?
18. 734756 — 340736 = ?	24. 875734 — 805345 = ?
19. 936475 — 463040 = ?	25. 730370 — 370370 = ?
20. 909909 — 98778 = ?	26. 100000 — 99999 = ?
21. 101010 — 90909 = ?	27. 100000 — 88888 = ?
22. 100000 — 77777 = ?	28. 666666 — 477777 = ?

29. From 67567 + 3456 take 9643 + 7345.
30. From 87567 + 2678 take 6304 + 3456.
31. From 73456 + 4345 take 9360 + 7561.
32. From 93464 + 7560 take 4234 + 961.
33. From 8345 + 6734 take 9641 + 1013.
34. From 99875 + 2634 take 7342 + 206.
35. From 8756 + 937 take 7309 + 561.
36. From 3456 + 9879 take 4050 + 345.

37. From one thousand one hundred and one, subtract nine hundred and eleven.

38. From fifty thousand, subtract five thousand five hundred and five.

39. From one million, subtract one hundred and one.

40. From thirty millions, thirty thousand and thirty, subtract three millions three thousand and three.

41. What time elapsed from the flood, 2348 A. C., to the death of Abraham, 1821 A. C.?

42. What time elapsed from the death of Abraham to the revolt of the ten tribes of Israel, 957 A. C.?

43. How many years from the first settlement in Greece, 1850 A. C., to the founding of Rome by Romulus, 753 A. C.?

44. How many years from the destruction of Troy, 1184 A. C., to the founding of Rome?

45. How many years from the discovery of America by Christopher Columbus, in 1492, to the declaration of American independence, in 1776?

46. George Washington died in 1799, and was 67 years old. In what year was he born?

47. Benjamin Franklin was born in 1706, and died in 1790. How old was he when he died?

48. Cotton was first planted in the United States in 1769. How many years since?

49. Glass windows were first used in England in 1180. How many years since?

50. Newspapers were first published in 1630. How many years since?

51. Quills were used for writing in 636. How many years since?

52. The first permanent settlement in Virginia was made in 1607. How many years since?

DIVISION.

SECTION VI.

30. Division is the process of finding how many times one number is contained in another.

31. The number divided is called the *dividend.* The number divided by is called the *divisor.* The result is called the *quotient.* When any thing remains after dividing, it is called the *remainder*, and is always of the same kind as the dividend.

Obs. The term *quotient* is from the Latin word *quoties*, signifying *how many times.*

32. The sign of division is a horizontal line between two dots, $\div$, and is read *divided by.* Thus $12 \div 3 = 4$ is read, 12 divided by 3 is equal to 4.

What is division? What is the dividend? What is the divisor? What is the quotient? What is the remainder? What is the sign of division?

33. When the divisor does not exceed 12, —

RULE. *Write the divisor at the left of the dividend. Find how many times the divisor is contained in the first left hand figure or figures, and write underneath the result. If there be no remainder, divide the next figure or figures in the same manner. If there be a remainder, suppose it to be prefixed to the next figure of the dividend, and divide as before.*

OBS. When the divisor is not contained in any figure of the dividend, excepting the first, a cipher must be written in the quotient.

1. Divide 8756 by 6.

Divisor, 6) 8756, dividend.
Quotient, 1459, and 2 remainder.

6 is contained in 8 once, and 2 over. Write the 1 underneath, in the quotient, and prefix the 2 to the next figure, 7, making 27. 6 in 27 4 times, and 3 over. Write the 4 in the quotient, and prefix the 3 to the 5, making 35. 6 in 35 5 times, and 5 over. Write the 5 in the quotient, and prefix the 5 to the next figure, 6, making 56. 6 in 56 9 times, and 2 over. Write the 9 in the quotient, and the 2 at the right for the remainder.

EXAMPLES.

2. Divide 845678 by 4.
3. Divide 967834 by 2.
4. Divide 603406 by 3.
5. Divide 734842 by 8.
6. Divide 356742 by 9.
7. Divide 498756 by 12.
8. Divide 643275 by 7.
9. Divide 734562 by 5.

34. When the divisor exceeds 12, —

RULE. *Write the divisor at the left of the dividend. Find how many times the divisor is contained in the smallest number of figures that will contain it one or more times, and write the result in the quo-*

Recite the rule for division when the divisor does not exceed 12. When the divisor exceeds 12, what is the rule?

tient at the right of the dividend. Multiply the divisor by this quotient figure, and subtract the product from the figures divided, and to the remainder annex the next figure of the dividend, and divide this number as before, and continue dividing in the same manner till all the figures are divided.

PROOF. *Multiply the divisor by the quotient, and to the product add the remainder, and if the sum be equal to the dividend, it is supposed to be right.*

OBS. 1. The dividend, divisor, and quotient must be separated by a line between them.

OBS. 2. If the remainder, after having one figure annexed, will not contain the divisor, write a cipher in the quotient, and annex another figure to the dividend.

OBS. 3. If the product of the divisor by the quotient figure be larger than the dividend, the quotient figure is too large.

OBS. 4. If the remainder, before a figure of the dividend has been annexed, be greater than the divisor, or equal to it, the last figure of the quotient is too small.

OBS. 5. Annexing a figure is placing it at the right of another figure; prefixing a figure is placing it at the left of another.

10. Divide 8756424 by 324.

```
324 ) 8756424 ( 27026
      648
      ----
      2276
      2268
      -----
         842
         648
         ----
         1944
         1944
         ----
```

The smallest number of figures that will contain the divisor is three, 875; in which the divisor is contained 2 times and 227 remainder, to which annex 6, the next figure, making 2276, which contains the divisor 7 times and 8 remainder, to which annex 4, the next figure of the dividend, making 84, which is less than the divisor; write a cipher in the quotient, and annex 2, the next figure of the dividend, making 842, which contains the divisor 2 times and 194 remainder; to which annex 4, the next figure, making 1944, which contains the divisor 6 times, without a remainder.

11.	4875674 ÷ 14 = ?	18.	9645045 ÷ 804 = ?
12.	5960345 ÷ 24 = ?	19.	8756454 ÷ 963 = ?
13.	6876454 ÷ 34 = ?	20.	8340341 ÷ 134 = ?
14.	7936273 ÷ 43 = ?	21.	473276 ÷ 912 = ?
15.	3732984 ÷ 61 = ?	22.	87345678 ÷ 125 = ?
16.	7345674 ÷ 75 = ?	23.	8930314 ÷ 635 = ?
17.	96034567 ÷ 83 = ?	24.	9181745 ÷ 225 = ?

35. When the divisor consists of two or more figures, products may be first formed of the divisor and the nine digits, which will enable the pupil to determine at once how many times the divisor is contained in any partial dividend.

OBS. The figures first selected to be divided, and those consisting of the remainders, with the several figures of the dividend annexed, are called the partial dividends.

25.

```
                        36 ) 96686964 ( 2685749
                             72
                             ---
36×1 =  36                   246
36×2 =  72                   216
36×3 = 108                   ---
36×4 = 144                    308
36×5 = 180                    288
36×6 = 216                    ---
36×7 = 252                     206
36×8 = 288                     180
36×9 = 324                     ---
                                269
                                252
                                ---
                                 176
                                 144
                                 ---
                                  324
                                  324
                                  ---
```

26. Divide 7684564 by 324.
27. Divide 7675439 by 273.
28. Divide 8432564 by 346.

29. Divide 6543742 by 295.

30. Divide 4526437 by 425.

By comparing the several products of the divisor and the partial dividends together, the pupil will discover how many times the divisor is contained in any partial dividend. Thus it will be seen that 36 is contained in 96, twice; in 246, 6 times; in 308, 8 times, &c.

36. When the divisor is a composite number, —

RULE. *Divide the dividend by one of the factors of the divisor, and the quotient thus obtained by the other.*

To find the true remainder when there are two factors, —

RULE. *Multiply the first divisor by the last remainder, and to the product add the first remainder, which will be the true remainder.*

When there are more than two factors, —

RULE. *Multiply the product of the first and second divisor by the last remainder, and the first divisor by the second remainder; to the sum of their products add the first remainder, which will be the true remainder.*

37. This rule depends upon the principle stated in art. 31, that the remainder, after division, is of the same kind as the dividend.

31. Divide 96599 by $84 = 7 \times 4 \times 3$.

7) 96599		$7 \times 4 \times 2 = 56$
4) 13799,	6, 1st rem.	$7 \times 3 = 21$
3) 3449,	3, 2d rem.	$6 = 6$
1149,	2, 3d rem.	83, true rem.

When the divisor is a composite number, what may be done? How do you find the remainder when there are two factors? More than two factors?

The first remainder, 6, is obviously so many units. The second dividend being so many 7ths, the second remainder, 3, is so many 7ths, and must be multiplied by 7. The third remainder, 2, being so many 28ths, must be multiplied by 28, or 7 × 4, to reduce it to units. The sum of all the remainders thus reduced will be the true remainder.

32. Divide 47516 by 35.	35. Divide 4275 by 32.
33. Divide 97234 by 27.	36. Divide 3654 by 42.
34. Divide 87544 by 20.	37. Divide 2743 by 18.

38. When there are ciphers at the right of the dividend, —

RULE. *Cut off as many figures from the right of the dividend as there are ciphers in the divisor. Divide the remaining figures of the dividend by the significant figures of the divisor, and to the remainder annex the figures cut off from the dividend for the true remainder.*

OBS. 1. If there be no remainder, the figures cut off will be the true remainder.

OBS. 2. If the divisor be 10, 100, 1000, &c., the figures cut off will be the remainder, and the remaining figures of the dividend will be the quotient.

38. Divide 4654 by 300.	41. Divide 6702 by 10.
39. Divide 3756 by 400.	42. Divide 4200 by 20.
40. Divide 2640 by 270.	43. Divide 1364 by 40.

GENERAL PRINCIPLES IN DIVISION.

39. As the product and one factor in division is given to find the other, it is evident that *multiplying the dividend is the same in effect as multiplying the quotient, and dividing the dividend the same as dividing the quotient.* Thus 48 divided by 6 gives 8 for

---

When there are ciphers at the right of the dividend, what is the rule? What effect is produced on the quotient by multiplying or dividing the dividend?

the quotient. Now, multiplying 48 by 2, 3, 4, &c., will make the quotient 2, 3, 4, &c., times larger.

40. It is also evident that *multiplying the divisor is the same as dividing the dividend, and dividing the divisor the same as multiplying the dividend.* Thus, having 48 for a dividend and 6 for a divisor, if the divisor be multiplied by 2, 3, &c., it will give the same quotient as if the dividend were divided by 2, 3, &c., and by dividing the divisor by 2, 3, &c., it will give the same quotient as if the dividend were multiplied by the same numbers. Thus $48 \div \overline{6\times2} = 4$, which is the same as $48 \div 2$, and the quotient $\div$ by 6. Art. 43.

41. *If the dividend and divisor be both divided or multiplied by the same number, the quotient will not be changed.* Thus 48 divided by 6 will give the same quotient as $\overline{48\times4}$ divided by $\overline{6\times4}$; and $\overline{48\div3}$ divided by $\overline{6\div3}$.

42. When the sum and difference of two numbers are given, *the smaller number may be found by subtracting the difference from the sum and dividing the remainder by 2. The larger number may be found by adding the difference to the smaller number.*

PRACTICAL QUESTIONS.

1. A merchant owes to one man 7361 dollars, to another 1969 dollars, to a third 2739 dollars; how much does he owe to all three?

2. A man bought a farm for 9375 dollars, but was obliged to sell it for 845 dollars less than he gave for it; what did he sell it for?

3. A farmer sells 354 cords of wood at 4 dollars a cord, and 37 tons of hay at 13 dollars a ton; how much does he receive for all?

4. A gentleman left in his will his estate, valued

What effect is produced by multiplying or dividing the divisor? What by multiplying both the dividend and divisor? When the sum and difference of two numbers are given, how are the two numbers found?

at 64331 dollars, as follows: 5630 dollars to his wife, 1245 dollars for charitable purposes, the remainder to be divided equally among his seven children; how much did each child receive?

5. The salary of the President of the United States is 25000 dollars a year; how much will he save in one year, if he spend 50 dollars a day?

6. A merchant bought a ship for 27342 dollars, and paid for cargo 37564 dollars; he sold the ship and cargo in California for 164000 dollars; how much did he gain, after deducting 2560 dollars for the expenses of the voyage?

7. If you were to count 2 every second for twelve hours a day, how long would it take to count a million?

8. What is the difference between 644×46 and $615433 \div 13$?

9. A gentleman left his estate, valued at 16596 dollars, to be divided between his wife and three children in the following manner: his wife was to have one third of the whole estate; the oldest child was to receive one third of what was left; the remainder was to be equally divided between the two youngest; what did each receive?

10. The sum of two numbers is 1496, one of which is 984; what is the other number?

11. If a man's income be 1650 dollars a year, and he expends 24 dollars a week, how much will he have left at the end of the year?

12. The difference between two numbers is 965, the greater is 1875; what is the less number?

13. A merchant owes 14560 dollars, which is less than what is due him by 9560 dollars; what is the sum due him?

14. The difference between two numbers is 564, and the less number is 896; what is the greater number?

15. The product of two numbers is 1296, and one of the numbers is 9; what is the other?

16. If the quotient be 345, and the divisor 64, what must the dividend be?

17. If the remainder of a sum in division be 20, the quotient 423, and the divisor the sum of the quotient and remainder, plus 19, what will be the dividend?

18. There are two numbers, the greater of which is 37 times 45, their difference 4 times 19; what is their sum and product?

19. If the dividend be 1728, and the quotient 24, what will be the divisor?

20. There are three numbers, whose continued product is 3456; one of the numbers is 12, another is 18; what is the other number?

21. The sum of two numbers is 1296, their difference is 144; what are the numbers?

22. The sum of three numbers is 640, the difference between the least and the greatest is 220, and the difference between the middle number and the sum is 460; what are the numbers?

23. A grocer bought a hogshead of molasses, containing 133 gallons, at 27 cents a gallon; but 19 gallons having leaked out, he sold the remainder at 39 cents a gallon; did he gain or lose, and how much?

24. What is the difference between 19 times 144 and 6732 divided by 17?

EXAMPLES IN THE PRECEDING RULES, WITH SIGNS.

The signs of addition, multiplication, subtraction, and division have already been explained, Art. 27.

43. A vinculum, ———, or parenthesis, is used to collect several quantities into one. Thus $\overline{5+4}\times 6$, or $(5+4)\times 6$, signifies that the sum of 5 and 4 is to be multiplied by 6, which is 54. Without the vinculum

What is a vinculum?

$5+4\times 6$, only 4 is multiplied by 6, and the 5 is added to the product, making 29. Also, $\overline{12-3}\times 3$, or $(12-3)\times 3$, signifies that the difference between 12 and 3 is to be multiplied by 3, which is 27. Without the vinculum, the product of 3 by 3 is subtracted from 12, which leaves 3.

25. $6+5\times 4=$?
26. $\overline{6+5}\times 4=$?
27. $7+\overline{6-3}\times 5=$?
28. $\overline{7+6-3}\times 5=$?
29. $24\times 6\times 9\div 3=$?
30. $\overline{36-12}\times 4\div 8=$?
31. $\overline{96\div 16}\times 5+4=$?
32. $\overline{84\times 6}+\overline{8\times 9}\div 12=$?
33. $\overline{64-14}\times \overline{17-12}\div 5=$?
34. $\overline{96\div 16}\times \overline{18-12}=$?
35. $\overline{144\times 4}\div \overline{4\times 6}=$?
36. $16-9+\overline{24\div 2}\times 3=$?
37. $\overline{4\times 8}-12\times 4\div 2=$?
38. $64\times 6-7+8+9=$?

39. Write the sum of the products of 8 into 9, and 7 into 4.

40. Write the difference of the products of 8 into 9, and 7 into 4.

41. Write the product of the sum of 8 plus 9, and 7 plus 4.

42. Write the product of the difference of 9 minus 8, and 7 minus 4.

43. Write 24 divided by the product of 4 into 2.

44. Write the quotient of 24 for a dividend, and 3 for a divisor.

45. Write the quotient of 24 divided by the product of 3 and 2.

THE DIVISIBILITY OF NUMBERS.

44. *The divisibility of a number is its property of being exactly divided by another.* Thus 36 is divisible by 2, 3, 4, 6, 9, 12, and 18, because each of these numbers will divide it without a remainder.

What is the divisibility of a number?

45. *Every number divisible by another is also divisible by each one of its factors.* Thus 48 is divisible by 12, also by 4 and 3, and 2 and 6, factors of 12.

46. *Every number divisible by another is called a multiple of that number. If it be divisible by two or more numbers, it is a common multiple of those numbers.* Thus 24 is the common multiple of 3, 4, 6, 8, and 12.

47. *The least number divisible by two or more numbers is the least common multiple of those numbers.* Thus 12 is the least common multiple of 3, 4, and 6.

48. *Every number that will divide exactly another number is called the measure of that number.* Thus 4 is the measure of 8.

49. *Every number that will divide exactly two or more numbers is their common measure.* Thus 6 is the common measure of 12 and 18.

50. *The greatest number that will divide exactly two or more numbers is their greatest common measure.* Thus 8 is the greatest common measure of 24 and 56.

51. *Every number that will divide exactly another number will also divide any multiple of that number.* Thus 6 will divide 12; it will also divide 24, 36, 48, &c.

52. *Every number that will divide exactly two other numbers will also divide their sum, their difference, and their product.* Thus 3 will divide 6 and 15; it will also divide $6+15=21$, $6\times15=90$, and $15-6=9$.

53. *Every number divisible by 2 is called an even number; when not divisible by 2, an odd number.*

54. *Every number is divisible by 3 when the sum*

What is a multiple? What is a common multiple? What is the least common multiple? What is the measure of a number? What is the common measure? What is the greatest common measure? What is an even number? What is an odd number? What numbers are divisible by 3?

of its figures, considered as units, is divisible by 3. Thus 321 is divisible by 3, since 3+2+1=6 is divisible by 3.

55. *Every number is divisible by* 4 *when its two right hand figures are divisible by* 4. Thus 112, 116, 128, are divisible by 4, since 12, 16, 28, are divisible by 4.

56. *Every number is divisible by* 5 *when its right hand figure is* 0 *or* 5. Thus 10, 15, 20, 25, are all divisible by 5.

57. *Every even number is divisible by* 6 *which is divisible by* 3. Thus 12, 18, 24, 30, are all divisible by 6.

58. *Every number is divisible by* 9 *when the sum of its figures, considered as units, is divisible by* 9. Thus 8172 is divisible by 9, since 8+1+7+2=18 is divisible by 9.

PRIME AND COMPOSITE NUMBERS.

59. Whole numbers are either *prime* or *composite.*

60. A *prime number* is one which cannot be formed by multiplying together any two or more whole numbers greater than a unit, as 1, 2, 3, 5, 7, 11, 13.

61. A *composite number* is one which may be formed by multiplying together two or more whole numbers greater than a unit, as 4, 6, 9, 8, 10, 12.

OBS. A prime number cannot be divided by another number, except itself and a unit.

62. Every composite number may be resolved into prime numbers, which are called *prime factors.*

63. To find the prime factors of any composite number,—

RULE. *Divide the given numbers in succession by*

What numbers are divisible by 4? What by 5? What by 6? What by 9? What are prime numbers? What are composite numbers? What are prime factors?

the least prime number greater than a unit that will divide them without a remainder. The last quotient and the several divisors will be all its prime factors.

64. The principle of this rule is evident from the fact that the divisors are all prime numbers, and the last quotient also must be a prime number, since it cannot be divided by any number but itself and a unit.

1. What are the prime factors of 96?

$96 \div 2 = 48$. $48 \div 2 = 24$. $24 \div 2 = 12$. $12 \div 2 = 6$. $6 \div 2 = 3$. $2 \times 2 \times 2 \times 2 \times 2 \times 3 = 96$.

The given number, 96, is divided in succession by 2, the least prime number. The several prime factors are 2, 2, 2, 2, 2, 3.

EXAMPLES.

What are the prime factors of the following numbers?

2.	12 and	14	9.	18 and	20	16.	24 and	27
3.	15 "	21	10.	28 "	30	17.	32 "	34
4.	33 "	35	11.	36 "	40	18.	38 "	42
5.	44 "	48	12.	45 "	50	19.	49 "	51
6.	54 "	56	13.	58 "	63	20.	64 "	72
7.	76 "	80	14.	84 "	88	21.	90 "	96
8.	99 "	102	15.	104 "	108	22.	112 "	115

23. What are the prime factors of 128 and 132?
24. What are the prime factors of 136 and 144?
25. What are the prime factors of 156 and 160?
26. What are the prime factors of 164 and 168?
27. What are the prime factors of 176 and 192?
28. What are the prime factors of 216 and 224?
29. What are the prime factors of 234 and 336?
30. What are the prime factors of 375 and 450?
31. What are the prime factors of 470 and 540?
32. What are the prime factors of 560 and 680?

A Table of Prime and Composite Numbers.

No.	Factors.	No.	Factors.	No.	Factors.
1	prime.	34	2.17	67	prime.
2	prime.	35	5.7	68	2.2.17
3	prime.	36	2.2.3.3	69	3.23
4	2.2	37	prime.	70	2.5.7
5	prime.	38	2.19	71	prime.
6	2.3	39	3.13	72	2.2.2.3.3
7	prime.	40	2.2.2.5	73	prime.
8	2.2.2	41	prime.	74	2.37
9	3.3	42	2.3.7	75	3.5.5
10	2.5	43	prime.	76	2.2.19
11	prime.	44	2.2.11	77	7.11
12	2.2.3	45	3.3.5	78	2.3.13
13	prime.	46	2.23	79	prime.
14	2.7	47	prime.	80	2.2.2.2.5
15	3.5	48	2.2.2.2.3	81	3.3.3.3
16	2.2.2.2	49	7.7	82	2.41
17	prime.	50	2.5.5	83	prime.
18	2.3.3	51	3.17	84	2.2.3.7
19	prime.	52	2.2.13	85	5.17
20	2.2.5	53	prime.	86	2.43
21	3.7	54	2.3.3.3	87	3.29
22	2.11	55	5.11	88	2.2.2.11
23	prime.	56	2.2.2.7	89	prime.
24	2.2.2.3	57	3.19	90	2.3.3.5
25	5.5	58	2.29	91	7.13
26	2.13	59	prime.	92	2.2.23
27	3.3.3	60	2.2.3.5	93	3.31
28	2.2.7	61	prime.	94	2.47
29	prime.	62	2.31	95	5.19
30	2.3.5	63	3.3.7	96	2.2.2.2.2.3
31	prime.	64	2.2.2.2.2.2	97	prime.
32	2.2.2.2.2	65	5.13	98	2.7.7
33	3.11	66	2.3.11	99	3.3.11

No.	Factors.	No.	Factors.	No.	Factors.
100	2.2.5.5	135	3.3.3.5	170	2.5.17
101	prime.	136	2.2.2.17	171	3.3.19
102	2.3.17	137	prime.	172	2.2.43
103	prime.	138	2.3.23	173	prime.
104	2.2.2.13	139	prime.	174	2.3.29
105	3.5.7	140	2.2.5.7	175	5.5.7
106	2.53	141	3.47	176	2.2.2.2.11
107	prime.	142	2.71	177	3.59
108	2.2.3.3.3	143	11.13	178	2.89
109	prime.	144	2.2.2.2.3.3	179	prime.
110	2.5.11	145	5.29	180	2.2.3.3.5
111	3.37	146	2.73	181	prime.
112	2.2.2.2.7	147	3.7.7	182	2.7.13
113	prime.	148	2.2.37	183	3.61
114	2.3.19	149	prime.	184	2.2.2.23
115	5.23	150	2.3.5.5	185	5.37
116	2.2.29	151	prime.	186	2.3.31
117	3.3.13	152	2.2.2.19	187	11.17
118	2.59	153	3.3.17	188	2.2.47
119	7.17	154	2.7.11	189	3.3.3.7
120	2.2.2.3.5	155	5.31	190	2.5.19
121	11. 11	156	2.2.3.13	191	prime.
122	2.61	157	prime.	192	2.2.2.2.2.2.3
123	3.41	158	2.79	193	prime.
124	2.2.31	159	3.53	194	2.97
125	5.5.5	160	2.2.2.2.2.5	195	3.5.13
126	2.3.3.7	161	7.23	196	2.2.7.7
127	prime.	162	2.3.3.3.3	197	prime.
128	2.2.2.2.2.2.2	163	prime.	198	2.3.3.11
129	3.43	164	2.2.41	199	prime.
130	2.5.13	165	3.5.11	200	2.2.2.5.5
131	prime.	166	2.83	201	3.67
132	2.2.3.11	167	prime.	202	2.101
133	7.19	168	2.2.2.3.7	203	7.29
134	2.67	169	13.13	204	2.2.3.17

CANCELLATION.

65. Cancelling is the process of abridging arithmetical operations by striking out factors common to the divisor and dividend.

. **Rule.** *Write the numbers forming the dividend above a horizontal line, and the numbers forming the divisor below it; cancel all the factors common to the dividend and divisor. Then multiply and divide with the remaining figures.*

Obs. When any entire number is cancelled, either in the dividend or divisor, 1 must be written in its place.

66. This rule depends upon the principle illustrated in (Art. 41,) viz., that by multiplying or dividing the dividend and divisor by the same number, the quotient will not be changed.

1. Divide 24×6 by 12×3.

$$\frac{\overset{2}{\cancel{24}}\times\overset{2}{\cancel{6}}}{\underset{1}{\cancel{12}}\times\underset{1}{\cancel{3}}}=\frac{4}{1}=4$$

24 multiplied by 6 is divided by 12 multiplied by 3. Since 12 and 2 are factors of 24, cancel 12 in the divisor and in the dividend, and write the other factor, 2, over the 24. Since 3 is a factor of the divisor, and also of 6 in the dividend, cancel 3 in the divisor and dividend, and write 2, the other factor, over the 6. The product of the remaining figures, 2 into 2, will be 4, which being divided by 1, will remain unchanged.

EXAMPLES.

2. Divide 48×15 by 16×3.
3. Divide 72×8 by 12×9.

What is cancelling? Recite the rule. Upon what does the rule depend?

4. Divide 144×9 by 24×18.
5. Divide 342×8 by $42\times 6\times 4$.
6. Divide 396×12 by $18\times 6\times 4$.
7. Divide 484×16 by $8\times 6\times 12$.
8. Divide 724×4 by $8\times 4\times 6$.
9. Divide 365×5 by $5\times 5\times 5$.
10. Divide 426×6 by $4\times 3\times 8$.
11. Divide 828×9 by $18\times 9\times 4$.
12. Divide 936×24 by 36×9.
13. Divide $96\times 9\times 8$ by 8×12.

THE GREATEST COMMON MEASURE.

67. The greatest common measure of two or more numbers has been shown to be (Art. 50) the greatest number that will divide them without a remainder.

68. To find the greatest common measure of two or more numbers, —

When the numbers are small, the greatest common measure may readily be determined by inspection, or by dividing the numbers by the largest number that will divide them without a remainder. Thus the greatest common measure of 12 and 16 is evidently 4.

Obs. When there are more than two numbers, first find the greatest common measure of any two of them, then of this common measure and one of the other numbers. Proceed thus with all the numbers.

What is the greatest common measure or factor of the following numbers?

12 and 18.	14 and 35.	15 and 20.	18 and 27.
21 and 28.	24 and 30.	16 and 28.	27 and 81.
39 and 52.	28 and 63.	32 and 48.	35 and 63.
42 and 77.	56 and 84.	28 and 126.	56 and 84.
21 and 126.	26 and 117.	42 and 112.	32 and 144.

69. When the numbers are large, —

What is the greatest common measure? How may it be found when the numbers are small?

Rule. *Divide in succession the greater number by the less, and that divisor by the last remainder, till nothing remains. The last divisor will be the greatest common measure.*

70. This rule depends upon the principle illustrated in (Art. 52,) viz., that any number that measures any other number will also measure their *sum*, their *difference*, and any *multiple* of those numbers.

1. What is the greatest common measure of 720 and 612?

```
612 ) 720 ( 1
      612
     ------------------
      108 ) 612 ( 5
            540
           ------------------
             72 ) 108 ( 1
                   72
                  ---------------
                   36 ) 72 ( 2
                        72
                        --
```

The greatest number being divided by the less, and the last remainder by the last divisor, gives 36 as the greatest common measure. Since 36 measures 72, it will also measure 108, the sum of 36 and 72; it will also measure 612, a multiple of 36.

EXAMPLES.

2. What is the greatest common measure of 252 and 348?
3. What is the greatest common measure of 493 and 899?
4. What is the greatest common measure of 208 and 648?

Recite the rule, when the numbers are large.

5. What is the greatest common measure of 825 and 960?

6. What is the greatest common measure of 5184 and 6912?

7. What is the greatest common measure of 3242 and 7564?

8. What is the greatest common measure of 972 and 1468?

9. What is the greatest common measure of 656, 864, and 976?

10. What is the greatest common measure of 896, 764, and 938?

11. What is the greatest common measure of 372, 964, and 704?

THE LEAST COMMON MULTIPLE.

71. The least number divisible by two or more numbers is the *least common multiple* of those numbers, (Art. 47.)

72. To find the least common multiple of two or more numbers, —

Rule. *First cancel every number that will measure any other given number. Then cancel the largest factor common to the two largest numbers, and all the factors of the other numbers that are contained in the uncancelled factors. The continued product of the remaining figures will be the least common multiple.*

Obs. 1. The numbers should be written in order in a horizontal line, beginning with the least on the left, and ending with the largest on the right.

Obs. 2. The largest common measure and the least common multiple should not be confounded. The largest common measure is the largest number that will divide two or more numbers. The least common multiple is the least number that is divisible by two or more numbers.

73. It is evident that the least common multiple of two numbers is the product of the numbers after striking

What is the least common multiple? What is the rule?

out their greatest common divisor, or the largest factor common to them, as it is necessary that a number be contained in the product as a factor only once.

The principle is the same for finding the least common multiple for three or more numbers.

1. What is the least common multiple of 6, 8, 12, 16, 20, 24?

~~6~~, ~~8~~, ~~12~~, ~~16~~, ~~20~~, 24.
$2\times5\times24=240.$

First cancel 6, 8, 12, since they measure 24. Then cancel 4 in 20, since it is the largest factor common to 20 and 24, and write underneath the other factor, 5. Cancel 8 in 16, as it is the largest factor contained in 24. The other factors, $2\times5\times24=240$, the least common multiple.

2. What is the least common multiple of 14, 18, 21, 27, 28, 126?

~~14~~, ~~18~~, ~~21~~, ~~27~~, ~~28~~, 126.
$3\times2\times126=756.$

Cancel 14, 18, 21, since they measure 126. Cancel also 14 in 28, the largest factor common to 28 and 126, and also 9 in 27, the largest factor common to the uncancelled factors in 2 and 126. The product of $3\times2\times126=756$, the least common multiple.

EXAMPLES.

3. What is the least common multiple of 4, 6, 8, and 10?

4. What is the least common multiple of 3, 4, 7, and 14?

5. What is the least common multiple of 5, 7, 10, and 21?

6. What is the least common multiple of 6, 9, 18, and 24?

Upon what principle does the rule depend?

7. What is the least common multiple of 8, 12, 16, and 20?

8. What is the least common multiple of 10, 12, 16, 18, and 20?

9. What is the least common multiple of 2, 3, 4, 5, 6, 7, 8, 9?

10. What is the least common multiple of 18, 36, 72, and 108?

11. What is the least common multiple of 14, 21, 28, and 35?

12. What is the least common multiple of 15, 20, 35, 45, and 60?

13. What is the least common multiple of 23, 46, 69, and 92?

14. What is the least common multiple of 34, 51, 68, and 85?

15. What is the least common multiple of 37, 74, 111, and 148?

16. What is the least common multiple of 41, 82, 123, and 164?

17. What is the least common multiple of 25, 60, 75, and 100?

18. What is the least common multiple of 30, 70, 90, and 140?

19. What is the least common multiple of 65, 75, 95, and 125?

20. What is the least common multiple of 136, 144, and 284?

21. What is the least common multiple of 272, 384, and 756?

22. What is the least common multiple of 320, 450, and 640?

23. What is the least common multiple of 426, 560, and 846?

24. What is the least common multiple of 576, 646, and 736?

FRACTIONS.

SECTION VII.

74. A FRACTION is one or more equal parts of a unit.

75. Fractions are of two kinds, *common* and *decimal.*

A *common* fraction is composed of two terms, one written above the other, with a line between them. Thus, $\frac{4}{5}$, $\frac{3}{7}$. The term below the line is called the *denominator*, and shows the number of equal parts into which the unit is divided. The term above the line is called the *numerator*, and shows how many parts are expressed by the fraction.

OBS. The term *numerator* is from a Latin word, signifying to *number;* the *denominator* from a Latin word, signifying to *name.*

76. A fraction may also be considered as the quotient resulting from division, the numerator being the dividend, and the denominator the divisor. Thus $\frac{4}{7}$ is the quotient resulting from 4 being divided by 7, and may be read one seventh of four, or four sevenths of one. In the same manner,

$\frac{2}{3}$ is read one third of two, or two thirds of one.

$\frac{5}{8}$ is read one eighth of five, or five eighths of one.

$\frac{9}{7}$ is read one seventh of nine, or nine sevenths of one.

77. A *proper* fraction is one whose numerator is *less* than its denominator, as $\frac{2}{3}$, $\frac{3}{4}$, $\frac{5}{7}$.

78. An *improper* fraction is one whose numerator is *equal to* or *greater* than its denominator, as $\frac{4}{4}$, $\frac{9}{7}$, $\frac{6}{5}$.

OBS. Proper and improper fractions are also called *simple fractions.*

What are fractions? How many kinds of fractions? Of what is a fraction composed? How are fractions written? What are the terms of a fraction called? What does the term below the line show? What does the term above the line show? How may a fraction be considered? How are fractions read? What is a proper fraction? An improper fraction? A simple fraction?

79. A *mixed number* consists of a whole number and a fraction, as $5\frac{1}{2}$, $30\frac{1}{4}$, $16\frac{1}{2}$.

80. A *compound fraction* is a fraction of a fraction, or any number of fractions connected by the word *of*. Thus $\frac{3}{5}$ of $\frac{4}{7}$ of $\frac{4}{9}$.

81. A *complex fraction* is one which has a fraction or mixed number for its numerator or denominator, or both.

OBS. Any whole number may be considered as the fraction of another. Thus 4 is four twelfths of 12, 5 is five sevenths of 7.

82. *A fraction is multiplied by any number by multiplying its numerator or dividing its denominator by that number.*

83. *A fraction is divided by any number by dividing its numerator or multiplying its denominator by that number.*

84. *The value of a fraction is not changed by multiplying or dividing both of its terms by the same number.* Thus $\frac{3}{4}$ is equal in value to $\frac{6}{8}$, $\frac{9}{12}$, $\frac{15}{20}$, &c.; $\frac{6}{12}$ is also equal to $\frac{1}{2}$.

85. To reduce a fraction to its lowest terms, —

RULE. *Divide both terms by any numbers in succession that will divide them both without a remainder. Or divide both terms by their greatest common measure.*

1. Reduce $\frac{9}{36}$ to its lowest terms.

$3\,)\,\frac{9}{36}=\frac{3}{12} \quad 3\,)\,\frac{3}{12}=\frac{1}{4} \qquad 9\,)\,\frac{9}{36}=\frac{1}{4}$

Dividing both terms by 3 twice gives $\frac{1}{4}$, whose terms are the lowest. The same result is obtained by dividing both terms by 9, their greatest common measure.

What is a mixed number? A compound fraction? A complex fraction?

OBS. 1. A fraction is reduced to its lowest terms when no number greater than a unit will divide both of its terms without a remainder.

OBS. 2. The value of a fraction is not changed by reducing it to its lowest terms, (ART. 84.)

EXAMPLES.

2. Reduce $\frac{26}{39}$ to its lowest terms.

3. Reduce $\frac{11}{55}$.
4. Reduce $\frac{42}{64}$.
5. Reduce $\frac{56}{63}$.
6. Reduce $\frac{64}{96}$.
7. Reduce $\frac{96}{112}$.
8. Reduce $\frac{108}{112}$.
9. Reduce $\frac{110}{116}$.
10. Reduce $\frac{118}{196}$.
11. Reduce $\frac{124}{164}$.
12. Reduce $\frac{236}{484}$.
13. Reduce $\frac{576}{968}$.
14. Reduce $\frac{974}{1864}$.
15. Reduce $\frac{1236}{4764}$.
16. Reduce $\frac{2346}{8648}$.
17. Reduce $\frac{3366}{9674}$.
18. Reduce $\frac{4575}{9731}$.

86. To reduce an improper fraction to an equivalent whole or mixed number,—

RULE. *Divide the numerator by the denominator, and write the remainder, if there be any, over the denominator.*

This rule depends upon the fact stated in ART. 76, that the denominator of a fraction is the divisor, and the numerator the dividend.

19. Reduce $\frac{29}{5}$ to a whole or mixed number.

$$\begin{array}{r} 5\,)\,29 \\ \hline 5\frac{4}{5} \end{array}$$

Dividing the numerator, 29, by 5, its denominator gives $5\frac{4}{5}$, a mixed number.

What is the rule for reducing an improper fraction to a whole or mixed number?

Reduce the following examples to whole or mixed numbers: —

20. Reduce $\frac{27}{4}$.	26. Reduce $\frac{76}{8}$.
21. Reduce $\frac{37}{5}$.	27. Reduce $\frac{91}{9}$.
22. Reduce $\frac{41}{6}$.	28. Reduce $\frac{108}{11}$.
23. Reduce $\frac{54}{8}$.	29. Reduce $\frac{116}{13}$.
24. Reduce $\frac{64}{7}$.	30. Reduce $\frac{251}{16}$.
25. Reduce $\frac{71}{5}$.	31. Reduce $\frac{4561}{232}$.

87. To reduce a mixed number to an equivalent improper fraction, —

RULE. *Multiply the whole number by the denominator of the fraction, and to the product add the numerator, and write the sum over the denominator.*

32. Reduce $7\frac{2}{3}$ to its equivalent improper fraction.

Multiplying the 7 by 3, and adding the 2, gives 23 for a numerator, which written over the denominator, 3, gives $\frac{23}{3}$ for the improper fraction.

This rule is the converse of the preceding one.

OBS. 1. *Converse* means performed in an opposite order.

OBS. 2. A whole number may be changed to an improper fraction by writing 1 under it for a denominator.

OBS. 3. A whole number may be changed to an improper fraction of a given denominator by multiplying the whole number by the proposed denominator for a numerator.

Reduce the following examples to their equivalent improper fractions.

33. Reduce $8\frac{4}{5}$.	37. Reduce $16\frac{1}{2}$.
34. Reduce $9\frac{6}{7}$.	38. Reduce $30\frac{1}{4}$.
35. Reduce $10\frac{1}{3}$.	39. Reduce $272\frac{1}{4}$.
36. Reduce $12\frac{5}{7}$.	40. Reduce $69\frac{5}{6}$.

What is the rule for reducing a mixed number to an improper fraction?

41. Reduce $734\frac{92}{100}$.
42. Reduce $6730\frac{45}{1012}$.
43. Reduce $90101\frac{9}{7894}$.
44. Reduce $701\frac{1}{108}$.
45. Reduce 5, 6, 7, 8, 9, 11, to improper fractions.
46. Reduce 36 to thirds, 42 to fourths, 54 to sixths.
47. Reduce 64 to sevenths, 95 to sixteenths, 112 to twelfths.

88. To reduce compound fractions to simple fractions, —

RULE. *Multiply the numerators together for a numerator, and the denominators for a denominator.*

OBS. 1. All factors common to the numerator and denominator should be cancelled before multiplying.

OBS. 2. If a part of a compound fraction be a mixed or whole number, it must first be reduced to an improper fraction.

48. Reduce $\frac{3}{4}$ of $\frac{5}{7}$ to a simple fraction.

$\frac{1}{4}$ of $\frac{5}{7}$ is evidently $\frac{5}{28}$; $\frac{3}{4}$ is 3 times $\frac{5}{28}$, which is $\frac{15}{28}$, the simple fraction.

49. Reduce $\frac{2}{3}$ of $\frac{5}{6}$ of $\frac{3}{4}$ to a simple fraction.

$$\frac{\not{2}}{\not{3}} \quad \frac{5}{\underset{3}{\not{6}}} \quad \frac{\not{3}}{4} = \frac{5}{12}$$

Cancelling the factors 2 and 3 in the numerators, and also 2 and 3 in the denominators, gives 5 for a numerator, and the product of 3 and 4 for a denominator, which is $\frac{5}{12}$, the simple fraction.

EXAMPLES.

50. Reduce $\frac{4}{5}$ of $\frac{5}{6}$ of $\frac{7}{8}$ to a simple fraction.
51. Reduce $\frac{3}{7}$ of $\frac{8}{9}$ of $\frac{14}{15}$ to a simple fraction.
52. Reduce $\frac{1}{2}$ of $\frac{2}{3}$ of $\frac{3}{4}$ of $\frac{2}{5}$ of $\frac{7}{8}$ to a simple fraction.
53. Reduce $\frac{8}{9}$ of $\frac{18}{25}$ of $\frac{5}{11}$ of $\frac{22}{23}$ to a simple fraction.

What is the rule for reducing compound fractions to simple fractions?

54. Reduce $\frac{7}{13}$ of $\frac{26}{27}$ of $\frac{9}{13}$ of $\frac{39}{45}$ to a simple fraction.
55. Reduce $\frac{4}{9}$ of $\frac{27}{28}$ of $\frac{11}{13}$ of $\frac{39}{54}$ to a simple fraction.
56. Reduce $4\frac{1}{3}$ of 3 of $\frac{7}{9}$ of 8 to a simple fraction.
57. What is $\frac{2}{5}$ of $\frac{7}{8}$ of $10\frac{1}{2}$?
58. What is $\frac{3}{7}$ of $\frac{11}{13}$ of $91\frac{1}{3}$?
59. What is $\frac{4}{7}$ of $\frac{9}{11}$ of $16\frac{1}{2}$?
60. What is $\frac{9}{13}$ of $\frac{39}{18}$ of $19\frac{1}{2}$?
61. What is $\frac{7}{6}$ of $\frac{8}{14}$ of 112?
62. What is $\frac{6}{7}$ of $\frac{21}{28}$ of 30?
63. What is $\frac{9}{10}$ of $\frac{20}{33}$ of 100?
64. What is $\frac{2}{9}$ of $\frac{12}{13}$ of 260?

89. To change fractions which have different denominators to other equivalent fractions which shall have a common denominator, —

Rule. *Multiply both terms of each fraction by the product of the denominators of the other fractions. Or find the least common multiple of the denominators for a common denominator,* (Art. 71.) *Then multiply each numerator by that number which denotes the number of times its denominator is contained in the common multiple.*

65. Reduce $\frac{1}{2}$, $\frac{3}{4}$, and $\frac{4}{5}$ to a common denominator.

$$\frac{1 \times 4 \times 5}{2 \times 4 \times 5} = \frac{20}{40}. \quad \frac{3 \times 5 \times 2}{4 \times 5 \times 2} = \frac{30}{40}. \quad \frac{4 \times 4 \times 2}{5 \times 4 \times 2} = \frac{32}{40}.$$

It is obvious that the value of the fractions has not been changed, since the numerators and denominators of each have been multiplied by the same number, (Art. 84.)

Obs. It is generally best to find the least common multiple of the denominators, as this reduces the fractions to their lowest terms.

Recite the rule for changing fractions to a common denominator.

EXAMPLES.

66. Reduce $\frac{2}{3}$, $\frac{3}{4}$, $\frac{4}{5}$ and $\frac{5}{6}$ to a common denominator.

67. Reduce $\frac{6}{7}$, $\frac{7}{8}$, $\frac{8}{9}$, and $\frac{9}{10}$ to a common denominator.

68. Reduce $\frac{11}{12}$, $\frac{12}{13}$, $\frac{13}{14}$, and $\frac{14}{15}$ to a common denominator.

69. Reduce $\frac{2}{9}$, $\frac{5}{12}$, $\frac{17}{48}$, and $\frac{37}{72}$ to a common denominator.

70. Reduce $\frac{1}{2}$, $\frac{3}{4}$, $\frac{8}{19}$, and $\frac{7}{24}$ to a common denominator.

71. Reduce $\frac{1}{3}$ of $\frac{4}{5}$, $\frac{6}{7}$ of 8, and $\frac{1}{9}$ of $16\frac{1}{2}$ to a common denominator.

72. Reduce $\frac{2}{7}$ of $\frac{9}{11}$, $\frac{9}{13}$ of $\frac{4}{9}$, and 9 to a common denominator.

73. Reduce $\frac{2}{5}$, $\frac{3}{7}$, $\frac{9}{11}$, and $\frac{6}{13}$ to a common denominator.

74. Reduce $\frac{4}{5}$ of $2\frac{1}{3}$, $\frac{9}{24}$, and $\frac{7}{19}$ to a common denominator.

75. Reduce $\frac{11}{13}$, $\frac{7}{26}$, and $19\frac{1}{2}$ to a common denominator.

76. Reduce $\frac{4}{9}$, $\frac{8}{13}$, $\frac{7}{4}$, and $\frac{9}{15}$ to a common denominator.

77. Reduce $\frac{6}{7}$ of 9 and $\frac{3}{5}$ of 24 to a common denominator.

78. Reduce $\frac{4}{9}$, $\frac{7}{23}$, $\frac{8}{30}$, and $\frac{9}{22}$ to a common denominator.

79. Reduce $\frac{11}{12}$, $\frac{6}{11}$, $\frac{13}{15}$, and $\frac{7}{9}$ to a common denominator.

80. Reduce $\frac{4}{13}$, $\frac{8}{11}$, $\frac{15}{17}$, and 19 to a common denominator.

81. Reduce $\frac{5}{6}$ of $7\frac{1}{5}$ and $\frac{3}{7}$ of $9\frac{1}{2}$ to a common denominator.

90. To reduce one fraction to another of the same value, having a given numerator,—

RULE. *Multiply both terms of the fraction by the proposed numerator, and divide both terms by the numerator of the given fraction.*

82. Reduce $\frac{7}{9}$ to a fraction with 11 for its numerator.

$$\frac{7 \times 11}{9 \times 11} = \frac{77}{99}. \quad \frac{77 \div 7}{99 \div 7} = \frac{11}{14\frac{1}{7}}$$

Since both terms are multiplied and divided by the same numbers, the value of the fraction is not changed.

EXAMPLES.

83. Reduce $\frac{9}{13}$ to a fraction with 15 for a numerator.
84. Reduce $\frac{11}{17}$ to a fraction with 21 for a numerator.
85. Reduce $\frac{7}{24}$ to a fraction with 31 for a numerator.
86. Reduce $\frac{19}{20}$ to a fraction with 21 for a numerator.
87. Reduce $\frac{29}{37}$ to a fraction with 41 for a numerator.

91. To reduce one fraction to another of the same value, having a given denominator,—

RULE. *Multiply both terms of the fraction by the proposed denominator, and divide both terms by the denominator of the given fraction.*

88. Reduce $\frac{7}{11}$ to a fraction with 13 for a denominator.

$$\frac{7 \times 13}{11 \times 13} = \frac{91}{143} \quad \frac{91 \div 11}{143 \div 11} = \frac{8\frac{3}{11}}{13}$$

Since both terms are multiplied and divided by the same numbers, the value of the fraction is not changed.

What is the rule for reducing one fraction to another of a given numerator? What for reducing one to a given denominator?

EXAMPLES.

89. Reduce $\frac{9}{11}$ to a fraction with 19 for a denominator.

90. Reduce $\frac{13}{17}$ to a fraction with 21 for a denominator.

91. Reduce $\frac{21}{29}$ to a fraction with 19 for a denominator.

92. Reduce $\frac{31}{19}$ to a fraction with 45 for a denominator.

93. Reduce $\frac{26}{9}$ to a fraction with 17 for a denominator.

94. Reduce $\frac{33}{75}$ to a fraction with 23 for a denominator.

95. Reduce $\frac{42}{83}$ to a fraction with 96 for a denominator.

ADDITION OF FRACTIONS.

92. Rule. *Reduce the fractions, when necessary, to their least common denominators; add their numerators, and write their sum over the common denominator.*

Obs. 1. All whole and mixed numbers must first be changed to improper fractions, and compound fractions to simple fractions.

Obs. 2. In mixed and whole numbers, the whole numbers and fractions may be added separately, and their sums united.

96. What is the sum of $\frac{1}{3}$, $\frac{1}{4}$, and $\frac{2}{5}$?

$$\frac{1\times4\times5}{3\times4\times5}=\frac{20}{60}.\quad\frac{1\times3\times5}{4\times3\times5}=\frac{15}{60}.\quad\frac{2\times4\times3}{5\times4\times3}=\frac{24}{60}.$$

$$\frac{20}{60}+\frac{15}{60}+\frac{24}{60}=\frac{59}{60}.$$

The fractions, reduced to a common denominator, are $\frac{20}{60}$, $\frac{15}{60}$, $\frac{24}{60}$; which being added together are $\frac{59}{60}$.

93. It is evident that parts of a number of the same kind can be added in the same manner as whole numbers of the same kind. Thus, one sixtieth added to one sixtieth is two sixtieths, and so of any number of sixtieths.

Recite the rule for the addition of fractions.

EXAMPLES.

97. What is the sum of $\frac{3}{4}$, $\frac{3}{4}$, and $\frac{2}{4}$?
98. What is the sum of $\frac{2}{6}$, $\frac{5}{6}$, and $\frac{4}{6}$?
99. What is the sum of $\frac{2}{3}$, $\frac{3}{4}$, and $\frac{1}{3}$?
100. What is the sum of $\frac{3}{5}$, $\frac{4}{6}$, and $\frac{5}{7}$?
101. What is the sum of $\frac{4}{7}$, $\frac{7}{9}$, and $\frac{9}{11}$?
102. What is the sum of $\frac{8}{9}$, $\frac{11}{13}$, and $\frac{12}{17}$?
103. What is the sum of $\frac{11}{19}$, $\frac{13}{21}$, and $\frac{15}{27}$?
104. What is the sum of $\frac{3}{4}$ of $\frac{5}{7}$ and $\frac{7}{13}$ of $\frac{4}{9}$?
105. What is the sum of $\frac{7}{8}$ of $\frac{11}{13}$ and $\frac{17}{19}$ of $\frac{1}{2}$?
106. What is the sum of $\frac{2}{7}$ of 4 and $\frac{7}{11}$ of 9?
107. What is the sum of $3\frac{1}{2}$, $9\frac{1}{2}$, $7\frac{1}{2}$, and 8?
108. What is the sum of $\frac{1}{3}$ of $5\frac{1}{2}$, and $\frac{6}{7}$ of $11\frac{1}{2}$?
109. What is the sum of $\frac{7}{8}$ of $\frac{9}{10}$ and $\frac{4}{7}$ of $25\frac{1}{2}$?
110. What is the sum of $\frac{11}{15}$, $\frac{16}{17}$, $\frac{18}{19}$, and $\frac{21}{30}$?
111. What is the sum of $\frac{1}{2}$, $\frac{2}{3}$, $\frac{3}{4}$, $\frac{4}{5}$, and $\frac{6}{7}$?
112. What is the sum of $21\frac{1}{2}$, $14\frac{5}{6}$, and $\frac{7}{9}$ of 40?
113. What is the sum of $\frac{7}{8}$, $\frac{9}{10}$, $\frac{11}{12}$, and $\frac{10}{13}$?
114. What is the sum of $\frac{4}{9}$, $\frac{13}{16}$, $\frac{19}{20}$, and 5?
115. What is the sum of $\frac{6}{7}$ of $10\frac{1}{2}$ and $\frac{8}{9}$ of $30\frac{1}{4}$?
116. What is the sum of $\frac{3}{7}$ of $9\frac{1}{2}$ and $\frac{5}{6}$ of $64\frac{1}{5}$?

MULTIPLICATION OF FRACTIONS.

94. Rule. *Multiply the numerators together for a numerator, and the denominators for a denominator.*

Obs. 1. Whole and mixed nnmbers must first be changed to improper fractions.

Obs. 2. Cancel all factors common to the numerators and denominators.

The principle of this rule is the same as that for the reduction of compound fractions.

What is the rule for the multiplication of fractions?

117. Multiply $\frac{5}{7}$ by $\frac{4}{9}$.

First multiply $\frac{5}{7}$ by 4, which is $\frac{5}{7}\times 4$; but since the multiplier is not 4, but $\frac{1}{9}$ of 4, the product is 9 times too large. This product must therefore be divided by 9. $\frac{5}{7}\times\frac{4}{9}=\frac{20}{63}$.

118. Multiply $\frac{6}{14}$ by $\frac{2}{3}$.

$$\frac{\overset{2}{\not 6}}{\underset{7}{\not 1\not 4}}\times\frac{\not 2}{\not 3}=\frac{2}{7}$$

Cancel the two factors common to the numerator and divisor; the product of the remaining figures is $\frac{2}{7}$.

EXAMPLES.

119. Multiply $\frac{3}{4}$ by $\frac{5}{6}$.
120. Multiply $\frac{4}{5}$ by $\frac{7}{8}$.
121. Multiply $\frac{11}{13}$ by $\frac{15}{17}$.
122. Multiply $5\frac{1}{2}$ by $5\frac{1}{2}$.
123. Multiply $16\frac{1}{2}$ by $16\frac{1}{2}$.
124. Multiply $30\frac{1}{4}$ by $30\frac{1}{4}$.
125. Multiply 25 by $2\frac{1}{3}$.
126. Multiply 24 by $6\frac{2}{3}$.
127. Multiply $7\frac{1}{4}$ by 49.
128. Multiply $62\frac{1}{2}$ by 75.
129. Multiply 764 by $\frac{1}{16}$.
130. Multiply 849 by $\frac{1}{18}$.
131. Multiply $\frac{7}{8}$ by $\frac{4}{14}$.
132. Multiply $\frac{9}{11}$ by $\frac{8}{13}$.
133. Multiply $\frac{19}{21}$ by $\frac{14}{57}$.
134. Multiply $3\frac{1}{3}$ by 20.
135. Multiply 36 by $7\frac{1}{6}$.
136. Multiply 42 by $9\frac{1}{7}$.
137. Multiply 75 by $\frac{1}{7}$.
138. Multiply 64 by $8\frac{1}{8}$.
139. Multiply $\frac{1}{20}$ by 40.
140. Multiply $\frac{1}{99}$ by 675.
141. Multiply 975 by $\frac{6}{7}$.
142. Multiply 964 by $1\frac{1}{2}$.

SUBTRACTION OF FRACTIONS.

95. Rule. *Reduce the fractions as in addition, subtract the less numerator from the greater, and write the difference over the common denominator.*

Obs. 1. All whole and mixed numbers must first be reduced to improper fractions, and compound fractions to simple ones.

What is the rule for the subtraction of fractions?

OBS. 2. In whole and mixed numbers, the whole numbers may be subtracted separately, and then their difference united to the fraction.

143. From $\frac{3}{4}$ take $\frac{2}{5}$.

$$\frac{3 \times 5}{4 \times 5} = \tfrac{15}{20}. \quad \frac{2 \times 4}{5 \times 4} = \tfrac{8}{20}. \quad \tfrac{15}{20} - \tfrac{8}{20} = \tfrac{7}{20}.$$

Change $\frac{3}{4}$ and $\frac{2}{5}$ to a common denominator, which is $\frac{15}{20}$ and $\frac{8}{20}$. 8 from 15 leaves 7, which, written over the common denominator, 20, gives $\frac{7}{20}$ for the answer.

It is obvious that parts of numbers of the same kind can be subtracted in the same manner as whole numbers.

EXAMPLES.

144. From $\frac{5}{6}$ take $\frac{2}{6}$.
145. From $\frac{4}{9}$ take $\frac{2}{9}$.
146. From $\frac{3}{5}$ take $\frac{2}{7}$.
147. From $\frac{5}{7}$ take $\frac{4}{9}$.
148. From $\frac{6}{11}$ take $\frac{5}{13}$.
149. From $7\frac{1}{2}$ take $\frac{6}{7}$.
150. From $2\frac{3}{4}$ take $1\frac{2}{3}$.
151. From $16\frac{1}{2}$ take $5\frac{1}{3}$.
152. From $27\frac{1}{2}$ take $9\frac{2}{3}$.
153. From $19\frac{3}{4}$ take $7\frac{8}{9}$.
154. From $33\frac{1}{3}$ take $29\frac{4}{7}$.
155. From 9 take $3\frac{1}{2}$.
156. From $16\frac{3}{7}$ take 11.
157. From $226\frac{1}{9}$ take $37\frac{9}{11}$.
158. From 297 take $9\frac{19}{27}$.
159. From 201 take $9\frac{33}{37}$.
160. From $\frac{2}{3}$ of 96 take $\frac{3}{4}$ of 36.
161. From $\frac{6}{9}$ of $87\frac{1}{2}$ take $\frac{7}{8}$ of 32.
162. From the sum of $\frac{3}{4}$, $\frac{7}{8}$, and $\frac{10}{13}$ take $\frac{2}{7}$ of $3\frac{5}{6}$.
163. From the sum of $37\frac{1}{2}$ and $65\frac{2}{3}$ take $\frac{11}{12}$ of $83\frac{2}{3}$.
164. From the sum of $336\frac{1}{2}$ and $21\frac{9}{10}$ take $\frac{13}{27}$ of $67\frac{1}{3}$.
165. From the sum of $973\frac{7}{4}$ and $26\frac{2}{7}$ take $275\frac{9}{3}$ and $26\frac{7}{3}$.

DIVISION OF FRACTIONS.

96. RULE. *Invert the divisor, and proceed as in multiplication.*

OBS. 1. The divisor is inverted by writing the denominator for a numerator and the numerator for a denominator.

What is the rule for the division of fractions?

OBS. 2. Whole and mixed numbers must first be changed to improper fractions, and compound fractions to simple fractions. Cancel when possible.

166. Divide $\frac{5}{7}$ by $\frac{4}{9}$.

Dividing $\frac{5}{7}$ by 4 gives $\frac{5}{7}\times\frac{1}{4}$; but since the divisor, 4, is divided by 9, and dividing the divisor is the same as multiplying the quotient, (ART. 40,) the quotient $\frac{5}{7}\times\frac{1}{4}$ must be multiplied by 9, which gives $\frac{5}{7}\times\frac{9}{4}=\frac{45}{28}$.

The principle of this rule has already been explained in ART. 83.

EXAMPLES.

167. Divide $\frac{7}{9}$ by $\frac{3}{4}$.
168. Divide $\frac{5}{7}$ by $\frac{4}{5}$.
169. Divide $\frac{9}{11}$ by $\frac{13}{14}$.
170. Divide $\frac{11}{15}$ by $\frac{19}{20}$.
171. Divide $\frac{19}{21}$ by $\frac{7}{11}$.
172. Divide 24 by $\frac{1}{7}$.
173. Divide $\frac{1}{9}$ by 36.
174. Divide $\frac{1}{50}$ by 780.
175. Divide 640 by $\frac{1}{301}$.
176. Divide 780 by $\frac{2}{147}$.
177. Divide $9\frac{1}{2}$ by $846\frac{1}{2}$.
178. Divide $30\frac{1}{4}$ by $5\frac{1}{2}$.
179. Divide $272\frac{1}{4}$ by $16\frac{1}{2}$.
180. Divide 242 by $69\frac{1}{2}$.
181. Divide 345 by $5\frac{1}{2}$.
182. Divide 176 by $272\frac{1}{4}$.
183. Divide 464 by $30\frac{1}{4}$.
184. Divide 45 by $2\frac{1}{4}$.
185. Divide 764 by $31\frac{1}{2}$.
186. Divide 841 by $1\frac{1}{2}$.
187. Divide 765 by $3\frac{1}{2}$.
188. Divide $\frac{1}{1001}$ by 849.

OBS. 1. The reciprocal of any number is 1 divided by that number. Thus the reciprocal of 4 is $\frac{1}{4}$.

OBS. 2. Dividing by any number is the same as multiplying by its reciprocal. Thus dividing 9 by 4 is $\frac{9}{4}$. Multiplying 9 by $\frac{1}{4}$ is the same, $\frac{9}{4}$.

97. A complex fraction is only one form of the division of one fraction by another fraction, (ART. 81.)

98. To reduce complex fractions to simple ones, —

RULE. *Prepare the fractions, and proceed as in the division of fractions.*

What is the rule for reducing complex fractions?

This rule depends upon the fact that the numerator of every fraction is a dividend and the denominator a divisor, (Art. 76.)

189. Reduce $\frac{\frac{1}{2}}{\frac{1}{3}}$ to a simple fraction.

$$\tfrac{1}{2}\times\tfrac{3}{1}=\tfrac{3}{2}$$

Dividing $\frac{1}{2}$ by $\frac{1}{3}$ gives $\frac{3}{2}$ for the simple fraction.

Reduce the following complex fractions to simple ones.

EXAMPLES.

190. Reduce $\frac{\frac{2}{3}}{\frac{4}{5}}$.

191. Reduce $\frac{\frac{7}{9}}{\frac{11}{13}}$.

192. Reduce $\frac{\frac{3}{8}}{\frac{11}{13}}$.

193. Reduce $\frac{\frac{4}{17}}{12}$.

194. Reduce $\frac{\frac{2}{7}}{6}$.

195. Reduce $\frac{\frac{9}{21}}{74}$.

196. Reduce $\frac{\frac{11}{19}}{240}$.

197. Reduce $\frac{\frac{8}{17}}{4}$.

198. Reduce $\frac{\frac{13}{4}}{9}$.

199. Reduce $\frac{\frac{21}{6}}{7}$.

200. Reduce $\frac{\frac{11}{19}}{27}$.

201. Reduce $\frac{\frac{7}{24}}{3}$.

202. Reduce $\frac{\frac{3}{4}}{\frac{7}{6}}$.

203. Reduce $\frac{\frac{11}{19}}{8}$.

99. Complex fractions, after being reduced to simple ones, may be *added*, *multiplied*, *subtracted*, and *divided* in the same manner as simple fractions.

Obs. Whenever in the reduction of fractions there are factors common to the dividend and divisor, they should always be cancelled.

How may complex fractions be added, &c.?

PRACTICAL QUESTIONS.

204. Add $\dfrac{12\frac{5}{7}}{19} + \dfrac{\frac{8}{9}}{\frac{7}{6}}$.

205. Add $\dfrac{9\frac{1}{7}}{6\frac{1}{9}} + \dfrac{\frac{4}{9}}{11}$.

206. Add $\dfrac{5\frac{1}{2}}{5\frac{1}{2}} + \dfrac{30\frac{1}{4}}{30\frac{1}{4}}$

207. Add $\dfrac{\frac{27}{4}}{9} + \dfrac{9}{7\frac{1}{9}}$.

208. Subtract $\dfrac{19}{\frac{4}{6}}$ from $\dfrac{\frac{27}{2}}{7\frac{1}{2}}$.

209. Divide $\dfrac{9\frac{2}{3}}{7\frac{11}{13}}$ by $\dfrac{\frac{17}{4}}{\frac{7}{9}}$.

210. Multiply $\dfrac{4\frac{1}{2}}{\frac{4}{3}}$ by $\dfrac{19}{\frac{7}{9}}$.

211. Multiply $\dfrac{27\frac{1}{2}}{\frac{9}{8}}$ by $\dfrac{21}{\frac{7}{8}}$.

212. Divide $\dfrac{\frac{7}{11}}{\frac{9}{12}}$ of $\dfrac{\frac{5}{9}}{\frac{8}{11}}$ by $\dfrac{\frac{19}{3}}{\frac{72}{11}}$ of $10\frac{1}{2}$.

213. Multiply $\frac{3}{7}$ of $\dfrac{8\frac{9}{11}}{6\frac{4}{9}}$ by $\frac{7}{9}$ of $\dfrac{21\frac{1}{2}}{34\frac{1}{3}}$.

214. Multiply $\dfrac{4\frac{1}{2}}{7\frac{1}{5}}$ of $\dfrac{9\frac{1}{7}}{11\frac{1}{13}}$ by $\dfrac{7\frac{1}{2}}{21}$ of $\dfrac{4\frac{1}{7}}{94}$.

215. Divide $\frac{19}{20}$ of $\dfrac{74}{83\frac{1}{2}}$ by $\dfrac{9\frac{1}{2}}{7\frac{1}{11}}$ of $\dfrac{6}{17\frac{1}{5}}$.

216. Divide $\dfrac{13\frac{1}{3}}{16\frac{4}{19}}$ of 7 by $\dfrac{8\frac{1}{8}}{121}$ of $13\frac{1}{3}$.

217. What is the sum of $\frac{5}{11}$, $\frac{4}{15}$, $\frac{6}{13}$, and $\frac{7}{31}$?

218. What is the sum of $\frac{3}{7}$ of $7\frac{1}{2}$ and $\frac{5}{9}$ of 24?

219. What is the product of $\frac{7}{9}$ of $\frac{11}{13}$ multiplied by $\frac{15}{17}$ of 29?

220. What is the product of $\frac{11}{17}$ of 9 multiplied by $\frac{13}{14}$ of $63\frac{1}{2}$?

221. What is the difference between $\frac{3}{8}$ of 8 and $\frac{3}{7}$ of 5?

222. If a bushel of oats cost $\frac{5}{7}$ of a dollar, what is of a bushel worth?

223. If $\frac{4}{11}$ of a yard of cloth is worth $\frac{9}{10}$ of a dollar, what is one yard worth?

224. A owns $\frac{1}{3}$ of a ship, B $\frac{1}{5}$, C $\frac{1}{7}$, and D the remainder. What part of the ship belongs to D?

225. What will $7\frac{3}{7}$ yards of cloth cost at $5\frac{3}{4}$ dollars per yard?

226. What will 137 barrels of flour cost at $6\frac{2}{5}$ dollars per barrel?

227. What will $654\frac{1}{3}$ feet of land cost at $9\frac{3}{7}$ cents per foot?

228. What will $\frac{19}{30}$ of an acre of land cost at 176 dollars per acre?

229. What will $137\frac{9}{11}$ tons of hay cost at $13\frac{2}{7}$ dollars per ton?

230. How many yards of cloth are there in three pieces, the first piece containing $19\frac{3}{4}$, the second $27\frac{3}{7}$, and the third $41\frac{3}{4}$ yards.

231. A boy gave $\frac{1}{2}$ of an orange to one companion, and $\frac{1}{4}$ of the remainder to another. How much did he keep for himself?

232. Add together the sum, difference, and product of $\frac{11}{13}$ and $\frac{9}{21}$.

233. If 25 dollars be paid for $2\frac{3}{7}$ tons of hay, what is the price of a ton?

234. A merchant sold $\frac{2}{3}$ of $\frac{1}{5}$ of a ship to one person, and $\frac{2}{7}$ of $\frac{11}{12}$ to another. What part of the whole ship had he left?

235. How much greater is the sum of $9\frac{4}{15}$ and $6\frac{3}{17}$ than their difference?

236. A gentleman bequeathed $\frac{3}{5}$ of his estate, which was worth 36000 dollars, to his son, $\frac{2}{7}$ of what remained to his daughter, and the rest to his widow. What sum *of money did each* receive?

237. If the circumference of a wheel of a locomotive be $7\frac{1}{3}$ feet, how many revolutions would it make between Boston and Providence, a distance of $41\frac{1}{2}$ miles, there being 5280 feet in a mile?

238. A lady's fortune was 8000 dollars, which was $\frac{1}{3}$ of $\frac{3}{5}$ of her brother's. How much was her brother's portion?

239. What number is that to which if you add $\frac{1}{2}$ of $\frac{2}{3}$ of $\frac{4}{5}$ the sum will be 1?

240. A grocer sold $23\frac{2}{3}$ pounds of tea for $1754\frac{9}{10}$ cents. How much was that a pound?

241. If one acre of land will produce $27\frac{2}{3}$ bushels of corn, how much will $19\frac{2}{3}$ acres produce?

242. A merchant sold $37\frac{2}{3}$ yards of cloth for $296\frac{3}{8}$ dollars. How much was that a yard?

243. If a barrel contain $2\frac{1}{8}$ bushels of apples, how many barrels would contain $96\frac{2}{3}$ bushels?

244. How many pounds of sugar, at $6\frac{1}{4}$ cents per pound, can be bought for $960\frac{1}{2}$ cents?

245. How many coats, containing $1\frac{7}{8}$ yards, can be made from $16\frac{3}{4}$ yards? How much cloth will be left?

246. A man paid $364\frac{3}{8}$ dollars for Mocha coffee, at $\frac{3}{20}$ of a dollar a pound. How much coffee did he buy?

247. A gentleman who owns $\frac{3}{5}$ of a copper mine sells $\frac{1}{8}$ of his share for 1640 dollars. What was the mine worth?

248. If the greater of two fractions be $\frac{12}{17}$, and their difference $\frac{7}{18}$, what is the smaller fraction?

249. If the smaller of two fractions be $\frac{17}{65}$, and their difference $\frac{9}{13}$, what is the larger fraction?

250. If the sum of two fractions be $\frac{64}{68}$, and their difference $\frac{1}{37}$, what are the two fractions? (See Art. 42.)

251. What number is that, from which if you take $\frac{1}{7}$ of itself, the remainder will be 15?

252. What number is that, to which if you add $\frac{1}{3}$ of $\frac{2}{5}$ of itself, the sum will be 21?

253. What number is that which, being divided by $\frac{3}{4}$, and multiplied by $\frac{4}{5}$, will give 20 for a product?

254. A merchant bought $\frac{7}{9}$ of a ship, and afterwards sold $\frac{3}{7}$ of his share. What part did he keep for himself?

255. How many barrels of flour can be purchased for $735\frac{5}{8}$ dollars, at $4\frac{7}{8}$ dollars per barrel?

256. If there are $5\frac{1}{2}$ yards in a rod, how many rods are there in $1765\frac{2}{5}$ yards?

257. If there are $16\frac{1}{2}$ feet in a rod, how many rods are there in $1836\frac{4}{7}$ feet?

258. In an academy there are 240 pupils; $\frac{2}{3}$ of them are studying arithmetic; $\frac{5}{6}$ of the remainder are studying grammar, and the rest are studying geography. What is the number in each study?

259. A gentleman left to his elder son $\frac{5}{8}$ of his property, to the younger $\frac{2}{3}$ of the remainder, and the rest to his wife. The elder son had 2400 dollars more than the younger. How much did the wife receive, and how much was left to the family?

260. Three persons enter into trade — A, B, and C. A advances $\frac{5}{4}$ as much as B, C advances $\frac{4}{5}$ as much as B, which is 1200 dollars. The loss sustained by them was $\frac{7}{20}$ of the sum advanced. How much will each have left?

DECIMAL FRACTIONS.

100. A *decimal fraction* is one or more parts of a unit, divided into *tenths*, *hundredths*, *thousandths*, &c.

Obs. *Decimal* is from the Latin word *decem*, signifying *ten*.

101. Instead of writing the denominator, a point is placed at the left of the numerator, called the *decimal point*, which separates the decimal from the whole number.

102. The denominator of a decimal fraction, when written, is 1, with as many ciphers annexed as there are figures in the numerator.

103. The first place at the right of the decimal point is the place of tenths, the second the place of hundredths, the third the place of thousandths, the fourth the place of ten thousandths, &c.

104. The same figure, in decimals, represents a value ten times greater in the first place than in the second, and ten times greater in the second place than in the third, &c., and by each removal of any figure to the next place on the right, its representative value is diminished ten times.

105. Ciphers are used to keep significant figures in their proper places, and are written where no value is to be expressed.

106. As the law of the increase in the representative value of figures from right to left is the same in decimals as in whole numbers, they may be written together by placing the decimal point between them.

What are decimal fractions? How are decimals written? What is the denominator of a decimal? What is the first place of decimals called? the second? the third? the fourth? &c. For what are ciphers used, and where are they written? What is the law of increase in decimals?

107. Decimals may be *added*, *multiplied*, *subtracted*, and *divided* in the same manner as whole numbers.

OBS. A whole number and decimal written together is called a *mixed number*.

108. *Annexing* ciphers to decimals does not change their value, as the significant figures retain the same places as before Thus, .5, .50, .500, are decimals of the same value, for $\frac{5}{10}=\frac{50}{100}=\frac{500}{1000}$.

109. *Prefixing* ciphers to decimals diminishes their value ten times for every cipher prefixed, for each cipher removes the significant figures one place to the right. Thus, .4, .04, .004, are respectively equal to $\frac{4}{10}$, $\frac{4}{100}$, $\frac{4}{1000}$.

OBS. Ciphers are *annexed* when they are placed on the right of the decimal; they are *prefixed* when placed on the left of the decimal, and on the right of the decimal point.

110. A decimal may be *multiplied* by 10, 100, 1000, &c., by removing the decimal point as many places towards the right as there are ciphers in the multiplier. If there be not as many places, make the number equal by annexing ciphers. Thus .244 multiplied by 10, 100, 1000, &c., is 2.44, 24.4, 244, 2440, &c.

111. A decimal may be *divided* by 10, 100, 1000, &c., by removing the decimal point as many places towards the left as there are ciphers in the divisor. If there be not as many, make the number equal by prefixing ciphers. Thus .244 divided by 10, 100, 1000, &c., is .0244, .00244, .000244.

What is annexing ciphers? What is prefixing ciphers? What is their effect? How may decimals be multipled by 10, &c.? How may they be divided by 10, &c.?

The names of the different orders of decimals, and their relation to whole numbers, may be learned from the following table : —

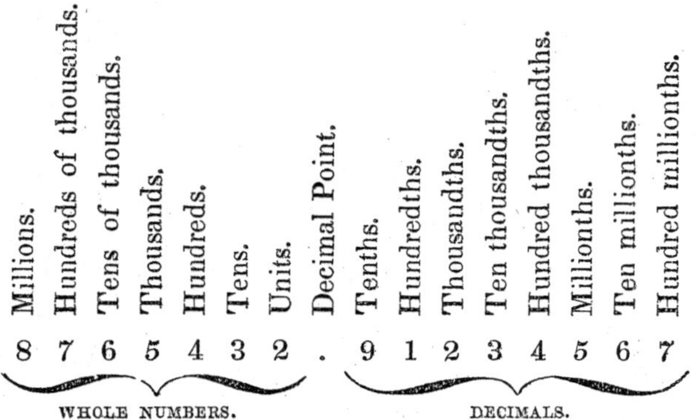

The whole numbers and decimals in the above table are expressed in words thus: Eight million, seven hundred and sixty-five thousand, four hundred and thirty-two, and ninety-one million, two hundred and thirty-four thousand, five hundred and sixty-seven hundred millionths.

112. To read decimals, begin at the first figure on the left of the decimal point, and read the figures as if they were whole numbers, and to the last add the name of its order. Thus: —

.5	is read 5 tenths.
.05	is read 5 hundredths.
.236	is read 236 thousandths.
.45678	is read 45678 hundred thousandths.
.365366	is read 365366 millionths.

Repeat the table. How are decimals read?

1. Read the following numbers: —

.4	.60	4.0101
.04	.060	2.0129
.004	.0600	16.2121
.0004	.6060	18.3004
.00004	.0606	42.0009
.000004	.00606	36.10009
.0000004	.000606	45.100009

Write the following numbers in figures: -

2. Seven hundred, and five tenths.
3. Seven, and one ten thousandth.
4. Seven hundred and five, and five millionths.
5. Two hundred and two, and two hundred thousandths.
6. Twenty-five million, and twenty-five ten millionths.
7. Three hundred and forty thousand, and three tenths.
8. Sixty-five thousand, and sixty-five ten thousandths.
9. Forty-four million, and forty-four ten millionths.
10. Five hundred thousand, and five hundred thousandths.
11. Sixty-four million, and sixty-four millionths.

REDUCTION OF DECIMALS.

113. To reduce a common fraction to a decimal, —

RULE. *Annex ciphers to the numerator, and divide by the denominator.*

OBS. The number of decimal places in the quotient will be equal to the number of ciphers annexed. When there are not as many, make the number equal by prefixing ciphers.

Recite the rule for reducing a common fraction to a decimal.

12. Reduce $\frac{3}{8}$ to a decimal.

$$8\ )\ \underline{3.000}$$
$$.375$$

As three ciphers were annexed to the numerator, three figures must be pointed off for decimals in the quotient.

This rule depends upon the principle stated in Art. 39, that if a number be multiplied and divided by the same number, its value remains unchanged.

13. Reduce the following fractions to decimals: —

$\frac{1}{4}$	$\frac{3}{4}$	$\frac{5}{2}$	$\frac{7}{16}$
$\frac{3}{7}$	$\frac{5}{4}$	$\frac{5}{9}$	$\frac{9}{16}$
$\frac{1}{5}$	$\frac{7}{8}$	$\frac{1}{16}$	$\frac{11}{16}$
$\frac{2}{5}$	$\frac{3}{8}$	$\frac{2}{16}$	$\frac{13}{16}$
$\frac{3}{5}$	$\frac{5}{8}$	$\frac{3}{16}$	$\frac{15}{16}$
$\frac{4}{5}$	$\frac{7}{8}$	$\frac{5}{16}$	$\frac{13}{20}$

114. Any common fraction can be expressed exactly in decimals whose denominator has no other factor but 2 and 5. If the denominator contains other factors, it cannot be expressed exactly in decimals.

115. To reduce a decimal fraction to a common fraction, —

Rule. *Write the denominator of the decimal under the numerator, and then reduce it to its lowest terms.*

14. Reduce .625 to a common fraction.

$$\frac{652}{1000} = \frac{5}{8}$$

The denominator of .625 is 1000. (Art. 102.) $\frac{625}{1000}$

What is the rule for reducing a decimal to a common fraction?

reduced to its lowest terms, (Art. 85,) gives $\frac{5}{8}$ for a common fraction.

EXAMPLES.

15. Reduce .125 to a common fraction.
16. Reduce .725 to a common fraction.
17. Reduce .925 to a common fraction.
18. Reduce .1125 to a common fraction.
19. Reduce .1265 to a common fraction.
20. Reduce .1475 to a common fraction.
21. Reduce .0045 to a common fraction.
22. Reduce .0208 to a common fraction.
23. Reduce .0046 to a common fraction.

ADDITION OF DECIMALS.

116. Rule. *Write the numbers of the same order under each other, so that the decimal points shall be in the same vertical column. Add as in whole numbers, and point off in the sum from the right as many places for decimals as are equal to the greatest number of decimal places in any of the given numbers.*

Proof. *Addition of decimals may be proved in the same manner as addition of whole numbers.*

The principle of this rule is the same as that in the addition of whole numbers, (Arts. 9, 17.)

24. Add together 3.014, 3014, 30.14, .003014, and .04045.

```
   3.014
3014.
  30.14
    .003014
    .04045
-----------
3047.197464
```

Write the numbers of the same order under each

What is the rule for the addition of decimals?

other, units under units, tens under tens, and tenths under tenths, hundredths under hundredths, &c., and add as in whole numbers, and point off from the right for decimals as many figures as are equal to the greatest number of decimal places in any of the given numbers. The greatest number of decimals in the above example is six; six figures must therefore be pointed off for decimals.

OBS. 1. The decimal point will always be exactly under the decimal points in the given numbers.

OBS. 2. The learner must be *very careful* in placing the decimal points according to the rule.

EXAMPLES.

25. Add 4.032, 64.5010, 96.081, 310.018, 1.0012.

26. Add 63.03036, 73.46030, .090345, 41.23101, 1.0109.

27. Add 340.14561, 84.960, .759112, . 000012, 2.0345.

28. Add 9.03456, 1.23456, 12.34567, 123.4567, 1234.567.

29. Add 85.05376, 5.45405, 54.04345, 540.4345, 7.00034.

30. 46.13455+9.73456+.0009345+875.+34.5=?

31. 9671.03+5.05674+8.7561+750.12+87.34=?

32. 1.45610+67563.1+2.31267+91.234+.679=?

33. Add together 1 tenth, 1 hundredth, 1 thousandth, and 1 ten thousandth.

34. What is the sum of 5 hundredths, 5 ten thousandths, and 5 millionths?

35. What is the sum of 95 millionths, 1 hundred thousandth, and 1 tenth?

36. What is the sum of 784 thousandths, 347 millionths, 75 ten thousandths, and 99 hundredths?

37. What is the sum of 465, 78 hundred thousandths, 9 millionths, 99 ten thousandths, and 9 tenths?

38. What is the sum of 475, 9 ten thousandths, 83 hundred thousandths, and 9 tenths?

MULTIPLICATION OF DECIMALS.

117. Rule. *Multiply as in whole numbers, and point off from the right of the product as many figures as there are decimal places in both factors.*

Proof. *Multiplication of decimals may be proved in the same manner as multiplication of whole numbers, or by changing them to the form of common fractions.*

Obs. If there be not as many decimal places in the product as in both factors, make the number equal by prefixing ciphers.

Multiply .0075 by .0025.

```
   .0075
   .0025
   -----
     375      75/10000 × 25/10000 = 1875/100000000.
    150
---------
.00001875
```

As there are eight decimal places in both factors, and only four in the product, four ciphers must therefore be prefixed to the product.

The reason of this rule will be evident by changing the decimal multiplicand and multiplier to the form of a common fraction, and then multiplying.

118. When the multiplier is a whole number, the multiplicand is taken as many times as there are units in the multiplier; but when the multiplier is a fraction, only such parts of the multiplicand are taken as are indicated by the multiplier. When the multiplier is less than a unit, the product will be less than the multiplicand.

What is the rule for the multiplication of decimals? When the multiplier is less than a unit, what is the product?

EXAMPLES.

Multiply the following decimals: —

39. 45.601 × 7.456.	46. 67.456 × 1.245.
40. 31.735 × 9.735.	47. 96.314 × 2.103.
41. 41.244 × 1.642.	48. 814.21 × 36.142.
42. 784.67 × 9.641.	49. 204.101 × 8.9614.
43. 46.043 × .0009.	50. 56.421 × 96.463.
44. 966.43 × .0061.	51. 42.001 × .00234.
45. 56.300 × .0312.	52. 84.241 × .00085.

53. Multiply 2 hundredths by 9 millionths.

54. Multiply 44 thousandths by 4 ten thousandths.

55. Multiply 1 ten millionth by 9 hundred thousandths.

56. Multiply 9999 by 9 hundred millionths.

57. Multiply 6 hundred millionths by 9 millionths.

SUBTRACTION OF DECIMALS.

119. Rule. *Write the less number under the greater, units under units, tenths under tenths, &c., so that the decimal points shall be exactly under each other. Subtract as in whole numbers, and point off from the right of the remainder as many places for decimals as are equal to the greatest number of places in either of the given numbers.*

Proof. *Subtraction of decimals may be proved in the same manner as subtraction of whole numbers.*

The principle of this rule is the same as that in subtraction of whole numbers.

58. From 961.345 take 2.456789.

```
961.345000
  2.456789
----------
958.888211
```

In this example three ciphers are annexed to the

What is the rule for the subtraction of decimals?

minuend, to make the number of decimals equal to the number in the subtrahend, which does not change the value of the minuend, (Art. 108.)

EXAMPLES.

59. From 103.013 take 95.0134.
60. From 96.401 take 65.12034.
61. From 965.14 take 9.45614.

62. 641.34—56.345.	65. 86.67—9.8675.
63. 7561.2—9.6456.	66. 101.1—90.9014.
64. 961.62—54.645.	67. 970.2—84.3456.

68. From 1 take 1 millionth.
69. From 9999 take 9 hundred thousandths.
70. From 5 take 555 ten millionths.
71. From 222 take 22 hundred thousandths.
72. From 99 and 9 hundredths take 9 ten thousandths.
73. From 1 million take 1 hundred thousandth.

DIVISION OF DECIMALS.

120. Rule. *Divide as in whole numbers, and point off from the right of the quotient as many figures for decimals as the decimal places in the dividend exceed those in the divisor; and if there be not as many, make the number equal by prefixing ciphers to the quotient.*

Proof. *Division of decimals may be proved in the same manner as division of whole numbers, or by common fractions.*

Obs. 1. When the decimal places in the divisor and dividend are *equal*, the quotient will be a *whole* number.

Obs. 2. When there are not as many decimal places in the dividend as in the divisor, ciphers may be annexed, and the division continued indefinitely. The ciphers thus annexed must be considered as decimal places.

What is the rule for the division of decimals?

OBS. 3. Unless great accuracy is required, it is not necessary to have more than five places of decimals in the quotient.

74. Divide .000288072 by 3.6.

```
3.6 ) .000288072 ( .00008002
       288
       ---
          072
           72
          ---
```

Since the dividend has nine decimal places, and the divisor but one, the quotient must have eight decimal places; four ciphers must therefore be prefixed to the quotient.

OBS. When there is a remainder after division, the sign + should be annexed to the quotient, to denote that the division may be continued farther.

EXAMPLES.

Divide the following decimals: —

75. 16.440÷.85.
76. 184.20÷9.6.
77. 2345.6÷.54.
78. 3.6751÷.04.
79. .0456÷.09.
80. 7645÷.90.
81. 47.58÷8.4.
82. 674.5÷.62.
83. 845.6701÷3.02.
84. 9.781234÷.246.
85. 45.67801÷.008.
86. 56789.1÷4.01.
87. 456.407÷.845.
88. 4702.41÷24.1.
89. 57.3745÷3.61.
90. 8.74562÷.096.

91. Divide one hundred and twenty-five, and nine ten thousandths, by six, and fifty-four ten thousandths.

92. Divide four hundred and eleven, and seven hundred and six millionths, by fifty-five, and ninety-three ten thousandths.

93. Divide seven million and one hundred and ten thousand, and ninety-four millionths, by eight hundred and forty-five ten thousandths.

94. Divide two hundred and twenty-four, and nine ten millionths, by three hundred and twenty hundred millionths.

PRACTICAL QUESTIONS.

95. What is the sum of 25.104, 25104, .6456, 45.52, 67.84, 96045.1, .8456, 75.104?

96. What is the sum of 4.6789, 567801.1, 5.0014, 375.27, 640.36, 8457.2?

97. What is the sum of 1 tenth, 1 hundredth, 1 thousandth, 1 ten thousandth, 1 hundred thousandth and 1 ten millionth?

98. Multiply six tenths by seven hundred ten thousandths.

99. Multiply four hundred, and four ten thousandths by ninety-seven millionths.

100. What cost nine tenths of a gallon of molasses, at three tenths of a dollar a gallon?

101. What cost 75.4 pounds of sugar, at .08 of a dollar a pound?

102. What is the difference between nine ten thousandths and nine hundred thousandths?

103. What is the difference between 444 thousandths and 4 one hundred millionths?

104. What is the difference between 9 and .8, and 6 and .08?

105. In one rod there are 16.5 feet. How many rods are there in 478964.54 feet?

106. Divide two hundred and fifty-six ten thousandths by four millionths.

107. If a pound of flour cost .05 of a dollar, how many pounds can be bought for 9650 dollars?

108. What would 1 bushel of oats cost, if 69.5 bushels cost 30.45 dollars?

109. How much hay can be purchased for 960.6 dollars, if 1 ton cost 12.8 dollars?

110. How many barrels of flour can be bought for 750.5 dollars, at 4.25 dollars a barrel?

111. How many gallons of molasses, at 33.5 cents per gallon, can be bought for 75.5 dollars?

6

CONTRACTIONS IN MULTIPLICATION AND DIVISION.

121. When the multiplier or divisor is an exact measure of 10, 100, 1000, &c., it may be changed to the form of a common fraction before multiplying or dividing, as in the following table: —

$3\frac{1}{3}=\frac{10}{3}$	$6\frac{1}{4}=\frac{100}{16}$
$8\frac{1}{3}=\frac{100}{12}$	$12\frac{1}{2}=\frac{100}{8}$
$16\frac{2}{3}=\frac{100}{6}$	$5=\frac{10}{2}$
$33\frac{1}{3}=\frac{100}{3}$	$25=\frac{100}{4}$
$333\frac{1}{3}=\frac{1000}{3}$	$125=\frac{1000}{8}$

122. To multiply by any of the numbers in the preceding table, —

RULE. *Annex as many ciphers to the multiplicand as there are ciphers in the numerator of the fraction, and divide by the denominator.*

EXAMPLES.

1. Multiply 4563 by $3\frac{1}{3}$.
2. Multiply 8844 by $8\frac{1}{3}$.
3. Multiply 9366 by $16\frac{2}{3}$.
4. Multiply 6372 by $33\frac{1}{3}$.
5. Multiply 8193 by $333\frac{1}{3}$.
6. Multiply 7348 by $6\frac{1}{4}$.
7. Multiply 8416 by $12\frac{1}{2}$.
8. Multiply 9264 by 5.
9. Multiply 6428 by 25.
10. Multiply 9416 by 125.

123. When one part of the multiplier is a multiple of the other, the process of multiplication may be shortened by the following rule: —

RULE. *Multiply by the unit figure, and then multiply this product by that number which denotes the number of times that this figure is contained in the two left hand figures of the multiplier.*

What is the rule for multiplying when the multiplier measures 10, 100, 1000, &c.? How may the process be shortened, when one part of the multiplier is a multiple of the other?

11. Multiply 46564 by 243.

```
    46564
      243
  -------
   139692     3 times the multiplicand.
  1117536   240 times the multiplicand.
  -------
 11315052
```

Multiply first by the unit figure, 3. Then, as 24, the remaining figures in the multiplier, is a multiple of 3, multiply the first product, 139692, by 8, which is 1117536. The unit figure of this product being written in the place of tens, makes the whole of this last product the same as 240 times the multiplicand. The sum of the partial products is 243 times the multiplicand.

EXAMPLES.

12. Multiply 428677 by 328.
13. Multiply 567894 by 427.
14. Multiply 678969 by 369.
15. Multiply 456784 by 728.
16. Multiply 734567 by 639.

124. To divide by any of the numbers in the preceding table,—

Rule. *Multiply by the denominator of the fraction, and point off from the right of the quotient for a remainder as many figures as there are ciphers in the numerator.*

EXAMPLES.

17. Divide 4675 by $3\frac{1}{3}$.
18. Divide 5789 by $8\frac{1}{3}$.
19. Divide 6542 by $16\frac{2}{3}$.
20. Divide 7345 by $33\frac{1}{3}$.
21. Divide 8456 by $333\frac{1}{3}$.
22. Divide 8796 by $6\frac{1}{4}$.
23. Divide 9234 by $12\frac{1}{2}$.
24. Divide 9541 by 5.
25. Divide 9645 by 25.
26. Divide 9846 by 125.

What is the rule for dividing when the divisor measures 10, &c.?

FEDERAL MONEY.

SECTION VIII.

125. The currency of the United States is styled *federal money.*

126. The denominations of federal money are *eagles, dollars, dimes, cents,* and *mills.*

127. The coins are made of gold, silver, and copper.

128. The gold coins are the *double eagle,* the *eagle,* the *half-eagle,* the *quarter-eagle,* and the *dollar.*

129. The silver coins are the *dollar,* the *half-dollar,* the *quarter-dollar,* the *dime,* and the *half-dime.*

130. The copper coins are the *cent* and the *half-cent.*

Obs. *Coin* is a piece of metal stamped with certain figures or characters by the authority of government.

TABLE OF FEDERAL MONEY.

10 mills	make	1	cent.
10 cents	"	1	dime.
10 dimes	"	1	dollar, ($.)
10 dollars	"	1	eagle.

It is now almost the universal practice in the United States to keep accounts in *dollars, cents,* and *mills.*

Obs. The mill is seldom regarded, except when great accuracy is required.

131. As the dollar is considered the unit or whole number, and cents and mills decimals, federal money may be *added, multiplied, subtracted,* and *divided,* in the same manner as decimals.

What is the currency of the United States styled? What are its denominations? What are the coins? Of what is each made?

REDUCTION OF FEDERAL MONEY.

132. RULE. *Dollars may be reduced to cents by annexing* TWO *ciphers.*

Dollars may be reduced to mills by annexing THREE *ciphers.*

Cents may be reduced to mills by annexing ONE *cipher.*

Dollars and cents may be reduced to cents by removing the decimal point to the left of the dollars.

Cents may be changed to dollars by pointing off the TWO *right hand figures.*

Mills may be changed to dollars by pointing off the THREE *right hand figures.*

Mills may be changed to cents by pointing off ONE *figure on the right.*

133. As pointing off *two* figures is the same as dividing by 100, and pointing off *one* figure the same as dividing by 10, the figures pointed off will be the same as the dividend, (ART. 31.)

EXAMPLES.

1. Reduce \$75 to cents.
2. Reduce \$96 to cents.
3. Reduce \$110 to mills.
4. Reduce \$87½ to mills.
5. Change 144 cents to dollars.
6. Change 1605 cents to dollars.
7. Change 6456 mills to dollars.
8. Change 1790 mills to dollars.

ADDITION OF FEDERAL MONEY.

134. RULE. *Write the numbers under each other, dollars under dollars, cents under cents, &c., and add as in whole numbers.*

PROOF. *The same as in addition of whole numbers.*

How are dollars reduced to cents? How to mills? How are cents reduced to mills? How are cents changed to dollars? How are mills changed to dollars? What is the rule for addition of federal money?

OBS. As cents occupy the places of tenths and hundredths, when there is but one figure expressing cents, a cipher must be prefixed to it.

EXAMPLES.

9. What is the sum of $84 and 15 cts., $97 and 9 cts., $137, 12 cts., and 9 mills?

10. What is the sum of $19, 6 cts., and 4 mills; $76, 7 cts., and 8 mills; $275, 10 cts., and 1 mill?

11. What is the sum of $1756, 4 cts., and 9 mills; $196, 7 cts., and 8 mills; $550, 5 cts., and 5 mills; $730, 17 cts., and 7 mills?

12. What is the sum of $1375, 10 cts., and 1 mill; $13, 75 cts., and 3 mills; $7, 3 cts., and 6 mills; $14, 8 cts., and 2 mills?

MULTIPLICATION OF FEDERAL MONEY.

135. RULE. *Multiply as in whole numbers, and point off from the right of the product as in multiplication of decimal fractions.*

PROOF. *The same as in multiplication of whole numbers.*

EXAMPLES.

13. What will 75 yards of cloth cost at $5.50 per yard?

14. What will 156 barrels of apples cost at $2.375 per barrel?

15. What will 344 pounds of cheese amount to at 6¼ cts. per pound?

16. What cost 436 bushels of wheat at $1¼ per bushel?

17. What cost 5750 feet of land at 7¾ cents per foot?

18. What cost 76 bushels of potatoes at 33⅓ cents per bushel?

19. What cost 86 acres of land at $45.375 per acre?

What is the rule for the multiplication of federal money?

SUBTRACTION OF FEDERAL MONEY.

136. Rule. *Write the numbers under each other, dollars under dollars, cents under cents, &c., and subtract as in whole numbers.*

Proof. *The same as in subtraction of whole numbers.*

EXAMPLES.

20. From $96.125 take $75.374.
21. From $116.013 take 99.101.
22. From $137, 6 cts., take $39, 7 cts., and 4 mills.
23. From $561, 9 cts., take $396, 14 cts., and 9 mills.
24. From $856, 10 cts., take $703, 9 cts., and 7 mills.
25. From $937, 5 cts., take $364, 15 cts., and 5 mills.

DIVISION OF FEDERAL MONEY.

137. Rule. *Divide as in whole numbers, and point off from the right of the quotient, as in division of decimal fractions.*

Proof. *The same as in division of whole numbers.*

EXAMPLES.

26. If 27 pounds of sugar cost $1.635, what will 1 pound cost?
27. If 56 barrels of flour cost $275.75, what will 1 barrel cost?
28. If 3242 feet of land cost $2825.75, what will 1 foot cost?
29. At $0.40 per yard, how many yards of calico can be bought for $15.56?

What is the rule for the subtraction of federal money? What is the rule for division?

30. At $0.37½ per yard, how many yards of gingham can be bought for $25.75?

31. How many times is $0.06 contained in $240.36?

32. How many times is $0.05 contained in $425.75?

33. How many times is $0.0875 contained in $58.645?

PRACTICAL QUESTIONS.

34. What is the sum of twenty-five dollars and nine cents; thirty-seven dollars, seven cents, and five mills; fifty-eight dollars, ten cents, and three mills?

35. What is the sum of one hundred and ten dollars, one cent, and nine mills; four hundred and forty dollars, six cents, and eight mills?

36. What cost 25.5 pounds of sugar at 12½ cts. per pound?

37. What cost 44¼ pounds of tea at 56¼ cts. per pound?

38. What cost 85½ pounds of beef at 7¼ cts. per pound?

39. What cost 4567 feet of land at 8¾ cts. per foot?

40. What is the difference between seven hundred and forty dollars, seven cents, and nine mills, and three hundred and thirty dollars, nine cents, and seven mills?

41. From one hundred dollars take nine cents and nine mills.

42. From one thousand dollars and ten cents take ninety dollars, nine cents, and nine mills.

43. How much must be added to four dollars, four cents, and four mills, to make five hundred dollars?

44. A man bought a farm for $7560.87, and sold it for $8050.50. How much did he gain?

45. If a man pay $375.50 for 85 barrels of flour, how much does he give a barrel?

46. If a man pay $42.50 for 27½ pounds of tea, how much does he give a pound?

47. If $425.50 were paid for 465½ bushels of corn, what would one bushel cost?

BILLS OF PARCELS.

138. A bill of parcels is an account given by the seller to the buyer of the articles purchased, with the price of each.

48. Boston, August 10, 1850.

Mr. James Blackington

Bought of William Jones

15 yards of broadcloth, at \$3.75 per yard,
12½ yards of cassimere, at \$2.37½ per yard,
7⅓ yards of satin, at \$4.12½ per yard,
5½ yards of linen cambric, at \$0.37½ per yard.

49. Boston, Sept. 1, 1850.

Mr. George Bright

Bought of J. D. Williams & Co.

4 boxes of sugar, of 240½ pounds each, at 6¼ cts. per pound,
61 bags of coffee, of 110½ pounds each, at 10½ cents per pound,
15 casks of rice, of 297½ pounds each, at 3¼ cts. per pound,
24 chests of tea, of 75½ pounds each, at 67½ cts. per pound,
19 barrels of Genesee flour, at \$5⅞ per barrel,

50. Boston, August 20, 1850.

Mr. Benjamin Southey

Bought of Wm. J. Reynolds & Co.

40 Webster's Dictionary, at \$4.75,
75 Worcester's History, at \$0.62½,
450 Colburn's Arithmetic, at \$0.12½,
84 Smellie's Philosophy, at \$0.67½,
230 Bibles, at \$0.87½,

DENOMINATE NUMBERS.

SECTION IX.

139. DENOMINATE NUMBERS express things of different kinds, or denominations, as 5 pounds, 6 shillings, 7 pence, &c.

TABLES OF MONEY, WEIGHTS, AND MEASURES.

English Money.

4 farthings (qr.)	=	1 penny,	d.
12 pence	=	1 shilling,	s.
20 shillings	=	1 pound,	£.
21 shillings	=	1 guinea.	

qr.		d.		s.		£.
4	=	1				
48	=	12	=	1		
960	=	240	=	20	=	1

OBS. A farthing is often expressed by $\frac{1}{4}$ d., 2 farthings by $\frac{1}{2}$ d., 3 farthings by $\frac{3}{4}$ d.

Troy Weight.

24 grains (gr.)	=	1 pennyweight,	dwt.
20 pennyweights	=	1 ounce,	oz.
12 ounces	=	1 pound,	lb.

gr.		dwt.		oz.		lb.
24	=	1				
480	=	20	=	1		
5760	=	240	=	12	=	1

OBS. 1. This weight is used in weighing gold, silver, and precious stones. It is also used in philosophical experiments.

OBS. 2. Troy weight may easily be changed into Avoirdupois, and Avoirdupois into Troy.

1 lb. Troy	=	$\frac{144}{175}$ lb. Avoirdupois.
1 oz. Troy	=	$\frac{192}{175}$ oz. Avoirdupois.

Avoirdupois Weight.

16 drams (dr.)	=	1 ounce, . . .	oz.
16 ounces	=	1 pound, . .	lb.
28 pounds	=	1 quarter, . . .	qr.
4 quarters	=	1 hundred weight,	cwt.
20 hundred weight	=	1 ton, ...	T.

dr.		oz.		lb.		qr.		cwt.		T.
16	=	1	=							
256	=	16	=	1						
7168	=	448	=	28	=	1				
28672	=	1792	=	112	=	4	=	1		
573440	=	35840	=	2240	=	80	=	20	=	1

Obs. By this weight all commodities are weighed except gold, silver, and precious stones.

Apothecaries' Weight.

20 grains (gr.)	=	1 scruple,	sc., or ℈.
3 scruples	=	1 dram,	dr., or ʒ.
8 drams	=	1 ounce,	oz., or ℥.
12 ounces	=	1 pound,	℔.

gr.		℈.		ʒ.		℥.		℔.
20	=	1						
60	=	3	=	1				
480	=	24	=	8	=	1		
5760	=	288	=	96	=	12	=	1

Obs. By this weight apothecaries mix their medicines.

Cloth Measure.

$2\frac{1}{4}$ inches (in.)	=	1 nail, ...	na.
4 nails	=	1 quarter of a yard,	qr.
4 quarters	=	1 yard, ...	yd.

in.		na.		qr.		yd.
$2\frac{1}{4}$	=	1				
9	=	4	=	1		
36	=	16	=	4	=	1

Obs. The Flemish ell is 3 quarters, the English ell is 5 quarters, and the French ell is 6 quarters.

Long Measure.

12 inches (in.)	=	1 foot, . . .	ft.
3 feet	=	1 yard, . . .	yd.
6 feet	=	1 fathom, . .	fath.
$5\frac{1}{2}$ yards, or $16\frac{1}{2}$ feet,	=	1 rod, . . .	r.
40 rods	=	1 furlong, . .	fur.
8 furlongs	=	1 mile, . . .	m.
3 miles	=	1 league, . .	lea.

in.	ft.	yd.	r.	fur.	m.	lea.
12 =	1					
36 =	3 =	1				
198 =	$16\frac{1}{2}$ =	$5\frac{1}{2}$ =	1			
7920 =	660 =	220 =	40 =	1		
63360 =	5280 =	1760 =	320 =	8 =	1	
190080 =	15840 =	5280 =	960 =	24 =	3 =	1

Obs. 1. Long measure is used in measuring distances, &c. The rod is sometimes called *perch*, or *pole*.

Obs. 2. The circumference of the earth is measured by degrees of latitude. 60 geographical miles, or $69\frac{1}{5}$ English miles, form 1 degree, and 360 degrees form the circumference.

Square Measure.

144 square inches (sq. in.)	=	1 square foot,	sq. ft.
9 square feet	=	1 square yard,	sq. yd.
$30\frac{1}{4}$ sq. yds., $272\frac{1}{4}$ sq. ft.,	=	1 square rod,	sq. r.
40 square rods	=	1 rood, . .	R.
4 roods	=	1 acre, . . .	ac.
640 acres	=	1 square mile,	sq. m.

sq. in.	sq. ft.	sq. yd.	sq. r.	R.	ac.
144 =	1				
1296 =	9 =	1			
39204 =	$272\frac{1}{4}$ =	$30\frac{1}{4}$ =	1		
1568160 =	10890 =	1210 =	40 =	1	
6272640 =	43560 =	4840 =	160 =	4 =	1

Obs. Square measure is used in measuring surfaces.

Cubic Measure.

1728 cubic inches (cub. in.) = 1 cubic foot, cub. ft.
27 cubic feet = 1 cubic yard, cub. yd.

16 cubic feet make one foot of wood, and 8 feet of wood one cord; or 128 cubic feet make one cord.

40 feet of round and 50 feet of hewn timber were formerly considered a ton, but timber is now chiefly sold by board measure.

cub. in.		cub. ft.		cub. yd.
1728	=	1		
46656	=	27	=	1

OBS. Cubic measure is used in measuring solid bodies.

Liquid Measure.

4 gills (gi.)	=	1 pint, . . .	pt.
2 pints	=	1 quart, . . .	qt.
4 quarts	=	1 gallon, . .	gal.
$31\frac{1}{2}$ gallons	=	1 barrel, . . .	bar.
63 gallons	=	1 hogshead, .	hhd.

gi.		pt.		qt.		gal.		bar.		hhd.
4	=	1								
8	=	2	=	1						
32	=	8	=	4	=	1				
1008	=	252	=	126	=	$31\frac{1}{2}$	=	1		
2016	=	504	=	252	=	63	=	2	=	1

OBS. 1. The English wine gallon contains 231 cubic inches. In some places, milk and malt liquors are sold by a measure which contains 282 cubic inches in a gallon.

OBS. 2. Only a few articles are bought and sold by the barrel. Hogsheads are used only in estimating the contents of cisterns, wells, and other large bodies of water. The terms *pipe* and *butt* are never used as exact measures of quantity, but simply designate casks of a certain shape or form.

OBS. 3. A gallon is divided by apothecaries into pints, fluid ounces, fluid drams, and minims. A pint of water is estimated at a pound.

Dry Measure.

2 pints (pt.)	=	1 quart, . .	qt.
8 quarts	=	1 peck, . . .	pk.
4 pecks	=	1 bushel, . .	bu.
36 bushels	=	1 chaldron, . .	ch.

pt.		qt.		pk.		bu.		ch
2	=	1						
16	=	8	=	1				
64	=	32	=	4	=	1		
2304	=	1152	=	144	=	36	=	1

Time.

60 seconds (sec.)	=	1 minute, .	min.
60 minutes	=	1 hour, . .	hr.
24 hours	=	1 day, . .	day.
7 days	=	1 week, . .	wk.

sec.		min.		hr.		day.		wk.
60	=	1						
3600	=	60	=	1				
86400	=	1440	=	24	=	1		
604800	=	10080	=	168	=	7	=	1

Obs. Leap years are those which are exactly divisible by 4. Thus 1840, 1844, &c., are leap years. A common month consists of 4 weeks.

Circular Measure.

60 seconds (″)	=	1 minute, . .	′
60 minutes	=	1 degree, .	°
30 degrees	=	1 sign, . . .	s.
12 signs, or 360 degrees,	=	1 circle, . .	cir.

″		′		°		s.		cir.
60	=	1						
3600	=	60	=	1				
108000	=	1800	=	30	=	1		
1296000	=	21600	=	360	=	12	=	1

REDUCTION OF DENOMINATE NUMBERS.

140. *Reduction* is the process of changing the denomination of any quantity without changing its value.

Thus 6 pounds being changed into shillings, becomes 120 shillings. This change is produced by multiplying 6 by 20, the number of shillings equal to a pound. 60 inches being changed to feet, becomes 5 feet. This change is produced by dividing 60 inches by 12, the number of inches equal to a foot.

141. To reduce denominate numbers from a higher to a lower denomination, —

RULE. *Multiply the highest denomination by that number which denotes how many units of the next lower denomination make* ONE *unit of the higher, and to the product add the next lower denomination. Proceed in this manner with each denomination.*

1. Reduce £14 12s. 9d. 2qr. to farthings.

```
 £.   s.  d.  qr.
 14   12  9   2
 20
-----
 292
  12
-----
 593
292
-----
3513
   4
-----
14054
```

14 pounds are reduced to shillings by multiplying by 20, because 20 shillings make one pound, and adding the 12 shillings makes 292 shillings. This being

What is reduction? What is the rule for reducing denominate numbers from a higher to a lower denomination?

multiplied by 12, because 12 pence make one shilling, and adding the 9 pence, makes 3513 pence. This being multiplied by 4, because 4 farthings make one penny, and adding the 2 farthings, makes 14054 farthings.

142. To reduce denominate numbers from a lower to a higher denomination, —

RULE. *Divide the given denomination by that number which denotes how many units of this denomination make one of the next higher. Proceed in this manner with each denomination.*

2. Reduce 14054 farthings to pounds.

```
 4 ) 14054
12 ) 3513, 2 rem.
20 ) 292, 9 rem.
       14, 12 rem.      £14, 12 s. 9 d. 2 qr.
```

Divide the farthings by 4, because 4 farthings make one penny. The quotient will be 3513 pence and 2 farthings over. Divide the pence by 12, because 12 pence make one shilling. The quotient will be 292 shillings and 9 pence over. Divide the shillings by 20, because 20 shillings make one pound. The quotient will be 14 pounds and 12 shillings over.

OBS. 1. Denominate numbers are changed from a higher to a lower by *multiplication*, from a lower to a higher by *division*. The former has been called *reduction descending*, the latter, *reduction ascending*.

OBS. 2. The addition and subtraction of denominate numbers is the same as the addition and subtraction of simple numbers, except in the subdivisions of the unit.

English Money.

3. In £20, 15 s. 9 d. 3 qr. how many farthings?
4. In £40, 10 s. 8 d. 2 qr. how many farthings?

What is the rule for reducing denominate numbers from a lower to a higher denomination?

5. In £240, 19s. how many pence?
6. In 144560 farthings how many pounds?
7. In 176842 farthings how many pounds?
8. In 6845 pence how many pounds?

Troy Weight.

9. In 10 lbs. 8 oz. how many grains?
10. In 9 lbs. 6 oz. 18 dwt. how many grains?
11. In 42645 grains how many pounds?
12. In 6456 pennyweights how many pounds?

Avoirdupois Weight.

13. In 4 qrs. 26 lbs. how many ounces?
14. In 2 cwt. 3 qrs. 20 lbs. how many drams?
15. In 5 tons, 2 cwt. 3 qrs. 6 lbs. how many drams?
16. In 7500 ounces how many hundred weight?
17. In 6456780 drams how many tons?

Cloth Measure.

18. In 45 yards how many nails?
19. In 65 yards how many inches?
20. In 45 yards how many inches?
21. In 244 inches how many yards?
22. In 614 inches how many yards?

Long Measure.

23. In 7 furlongs, 30 rods, 5 yards, how many feet?
24. In 2 miles, 6 furlongs, 14 rods, 12 feet, how many inches?
25. In 4 miles, 7 furlongs, 20 rods, 16 feet, how many inches?
26. In 324560 inches how many miles?
27. In 456740 inches how many miles?
28. In 675670 inches how many miles?
29. In 360 degrees how many inches?

Square Measure.

30. In 64 square rods and 20 square yards how many feet?

31. In 72 roods, 30 square rods, 30 square yards, how many inches?

32. In 2 acres, 3 roods, 25 square rods, 5 square yards, how many inches?

33. In 8756704 square inches how many acres?

34. In 9567805 square inches how many acres?

Cubic Measure.

35. In 56 cubic feet how many inches?

36. In 75 cubic feet how many inches?

37. In 14560 inches how many cubic feet?

38. In 234 cords of wood how many cubic feet?

39. In 27560 cubic feet of wood how many cords?

Dry Measure.

40. In 24 bushels, 3 pecks, how many quarts?

41. In 29 bushels, 2 pecks, how many pints?

42. In 3694 pints how many bushels?

43. In 4675 pints how many pecks?

Liquid Measure.

44. In 13 hhds. of wine how many pints?

45. In 42 hogsheads how many gills?

46. In 1345 pints how many hogsheads?

47. In 1464 gills how many hogsheads?

Time.

48. In 24 days, 10 hours, and 25 minutes, how many seconds?

49. In 365 days, 6 hours, how many seconds?

50. In 40345600 seconds how many months?

51. In 784560 minutes how many weeks?

Circular Measure.

52. In 50 degrees how many seconds?
53. In 6 signs, 10 degrees, how many seconds?
54. In 756700 seconds how many degrees?
55. In 967500 minutes how many signs?

ADDITION OF DENOMINATE NUMBERS.

143. RULE. *Write the numbers of the same denomination directly under each other. Add the numbers in the lowest denomination, and find, by division, how many units of the next higher denomination are contained in their sum. Write the remainder under the column or columns of the lowest denomination, and add the quotient to the column of the next higher denomination. Proceed thus to the end.*

PROOF. *The same as in addition of simple numbers.*

56.

£.	s.	d.	qr.
20	14	10	2
16	12	8	3
14	16	7	3
19	18	11	2
72	3	2	2

The sum of the column of farthings is 10 farthings, which is 2 pence and 2 farthings. Write the 2 farthings underneath, and add the 2 pence to the next column, whose sum is 38 pence, which is 3 shillings and 2 pence. Write the 2 pence underneath, and add the 3 shillings to the next column, whose sum is 63 shillings, which is 3 pounds and 3 shillings. Write the 3 shillings underneath, and add the 3 pounds to the next column, whose sum is 72, which write underneath.

What is the rule for the addition of denominate numbers?

EXAMPLES.

English Money.

57.				58.				59.			
£.	s.	d.	qr.	£.	s.	d.	qr.	£.	s.	d.	qr.
16	15	10	3	36	19	11	3	34	13	10	2
21	12	9	2	75	15	10	2	59	14	11	1
41	17	11	3	64	13	9	1	67	19	9	3
37	18	8	1	55	12	8	3	34	16	8	2

Troy Weight.

60.				61.				62.			
lb.	oz.	dwt.	gr.	lb.	oz.	dwt.	gr.	lb.	oz.	dwt.	gr.
12	10	16	23	16	10	15	18	25	10	16	22
13	9	18	20	15	9	19	17	27	11	13	20
14	11	17	19	17	8	18	12	30	9	14	16
20	10	15	17	19	11	16	21	42	8	12	15

Avoirdupois Weight.

63.						64.					
ton.	cwt.	qr.	lb.	oz.	dr.	ton.	cwt.	qr.	lb.	oz.	dr.
8	15	3	24	14	12	4	12	2	16	13	14
7	16	2	21	15	13	5	17	1	14	12	15
9	18	1	20	12	10	9	16	3	10	11	12
6	13	3	19	11	14	8	13	1	12	15	10

Cloth Measure.

65.			66.			67.		
yd.	qr.	na.	yd.	qr.	na.	yd.	qr.	na.
17	3	2	24	2	3	35	2	3
18	2	3	27	1	2	34	3	1
19	1	2	25	3	1	39	2	2
20	3	2	34	2	3	40	1	2

Long Measure.

68.

m.	fur.	rd.	yd.	ft.	in.
4	7	30	4	2	8
5	6	29	3	1	9
6	3	25	2	2	7
9	4	31	1	1	6

69.

m.	fur.	rd.	yd.	ft.	in.
11	6	24	3	1	9
13	7	26	4	2	8
16	5	33	2	1	7
18	4	37	3	2	11

Square Measure.

70.

ac.	R.	sq. r.	sq. yd.	sq. ft.
40	3	27	25	8
37	2	25	22	7
45	1	29	19	6
54	2	30	18	5

71.

ac.	R.	sq. r.	sq. yd.	sq. ft.
56	3	20	8	8
64	2	16	7	7
75	1	15	6	5
84	2	14	5	6

Cubic Measure.

72.

ft.	in.
139	1425
134	1634
135	1560
238	1140

73.

ft.	in.
116	1235
108	1475
116	1573
127.	1680

Liquid Measure.

74.

hhd.	gal.	qt.	pt.
43	50	3	1
22	45	1	0
33	46	3	1
22	34	2	0

75.

hhd.	gal.	qt.	pt.
42	60	3	1
23	54	2	0
13	40	1	1
12	36	3	1

Dry Measure.

76.

ch.	bu.	pk.	qt.	pt.
6	28	3	7	1
7	25	2	6	0
9	30	2	5	1
8	31	3	4	1

77.

ch.	bu.	pk.	qt.	pt
5	30	2	6	1
4	18	3	7	0
3	15	2	5	1
6	14	1	4	0

Time.

78.

y.	d.	h.	m.	s.
46	260	20	54	40
36	160	18	44	36
74	214	17	38	27
65	196	16	24	17

79.

y.	d.	h.	m.	s.
75	164	14	56	34
80	175	16	42	37
77	164	17	37	48
64	178	15	53	27

Circular Measure.

80.

s.	°	′	″
24	26	54	48
16	25	46	37
14	24	42	59
34	27	16	35

81.

s.	°	′	″
46	24	55	49
35	27	57	44
37	22	35	42
32	25	54	26

MULTIPLICATION OF DENOMINATE NUMBERS.

144. Rule. *Write the multiplier under the lowest denomination in the multiplicand. Multiply each denomination of the multiplicand, beginning with the lowest, and find how many units of the next higher denomination are contained in the product. Write underneath the remainder, and add the quotient to the product of the next higher denomination. Proceed thus to the end.*

What is the rule for the multiplication of denominate numbers?

PROOF. *The same as in multiplication of simple numbers.*

The principle of this rule is the same as that of multiplication of simple numbers.

OBS. When the multiplier is a composite number, multiply by each factor of the multiplier in succession.

82. Multiply £9 10s. 8d. 3qr. by 8.

£.	s.	d.	qr.
9	10	8	3
			8
76	5	10	0

8 times 3 farthings are 24 farthings, which are 6 pence, and no remainder. 8 times 8 pence are 64, and 6 added to the product make 70 pence, which are 5 shillings and 10 pence. Write the 10 underneath, and add the 5 to the next product, which is 80, making 85 shillings, which are 4 pounds and 5 shillings. Write the 5 underneath, and add the 4 to the next product which is 72, making 76, which write underneath.

EXAMPLES.

83.

£.	s.	d.	qr.
4	5	4	2
			2

84.

£.	s.	d.	qr.
8	7	11	3
			3

85.

£.	s.	d.	qr.
9	16	11	1
			4

86.

£.	s	d.
12	17	9½
		6

87.

£.	s.	d.
21	17	10¾
		8

88.

£.	s.	d.
19	12	8¼
		9

89.

£.	s.	d.
987	19	11¾
		12

90.

£.	s.	d.
945	17	8¼
		83

91.

£.	s.	d
966	13	9½
		42

92. What is the price of 27 lbs. of tea, at 5 s. $6\frac{1}{2}$ d. per lb.?

93. What is the price of 47 lbs. of sugar, at $8\frac{1}{2}$ d. per lb.?

94. What is the price of 63 tons of coal, at 13 s. 6 d. per ton?

95. What is the weight of 47 pieces of lead, each piece weighing 25 lbs. 6 oz. 12 dr.?

96. A farm consists of 9 fields, each 12 ac. 1 R. 32 rd. What is the extent of the farm?

97. What is the value of 576 lbs. of iron, at $3\frac{1}{2}$ d. per lb.?

98. What is the value of 15 pairs of shoes, at 7 s. 6d. per pair?

99. How much sugar in 15 boxes, each box containing 5 cwt. 3 qr. 15 lb.?

100. How much wood in 16 piles, each pile containing 10 C. 7 cub. ft. 15 cub. in.?

101. How many yards in 16 pieces of cloth, each piece containing 10 yds. 2 qr. 3 na.?

102. How many gallons of molasses in 75 pipes., each pipe containing 120 gal. 3 qt. 2 pt.?

103. How many bushels in 120 bbls. of potatoes, each barrel containing 2 bu. 1 pk.?

104. If a man travel 30 m. 7. fur. 35 rd. in 1 day, how far would he travel in 12 days?

105. If 1 acre produce 275 bu. 3 pk. of potatoes, how many bushels will 9 acres produce?

106. How many bushels of wheat in 84 sacks, each sack containing 3 bu. 2 pk. 1 qt.?

SUBTRACTION OF DENOMINATE NUMBERS.

145. RULE. *Write the less number under the greater, so that numbers of the same denomination shall be directly under each other. Begin to subtract with the*

What is the rule for the subtraction of deuominate numbers?

lowest denomination, and take each number in the lower line from the number above it, and write the remainder underneath. If the number in the lower line be greater than the number above it, add to the upper number as many units as make one of the next higher denomination, and then subtract, and add one to the next lower line before subtracting. Proceed thus to the end.

PROOF. *The same as in subtraction of simple numbers.*

107. From £20, 14s. 9d. 2qr. take £14, 15s. 10d. 3 qr.

£.	s.	d.	qr.
20	14	9	2
14	15	10	3
5	18	10	3

Beginning with the lowest denomination, subtract. As 3 farthings cannot be taken from 2, add to it 4 farthings, making 6; 3 from 6 leaves 3. Since 4 farthings were added to the farthings in the upper line, 1 penny, its equal, must be added to the 10 pence in the lower line, making 11 pence. As 11 cannot be taken from the 9, 12 pence must be added to the 9, making 21 pence; 11 from 21 leaves 10. As 12 pence were added to the upper line, 1 shilling, its equal, must be added to 15 in the lower line, making 16 shillings. As 16 cannot be taken from 14, 20 shillings must be added, making 34 shillings; 16 from 34 leaves 18. As 20 shillings were added to the upper line, 1 pound, its equal, must be added to the 14 in the lower line, making 15; 15 from 20 leaves 5.

EXAMPLES.

108. What is the difference between £3, 5 s. 6 d. and £2, 3 s. 2 d.?

109. What is the difference between £25, 12 s. 9 d. and £22, 13 s. 9 d.?

110. A merchant cuts 5 yds. 3 qr. 2 na. of cloth from a piece containing 21 yds. 2 qr. 2 na. How many yards remained?

111. The latitude of the Cape of Good Hope is 33° 56′ 13″ S., and that of Cape Horn 55° 58′ 40″ S. What is the difference of latitude between the two places?

112. The latitude of Boston is 42° 21′ 23″ N.; that of New Orleans 29° 8′ 32″ N. How much farther south is New Orleans than Boston?

113. The latitude of Paris is 48° 50′ 13″ N.; that of New York 40° 42′ 35″ N. How many degrees farther north is Paris than New York? How many degrees farther north is Paris than Boston?

114. The latitude of Quebec is 46° 49′ 12″. How many degrees farther north is Paris than Quebec?

115. The latitude of Charleston is 32° 46′ 33″ N.; that of Cincinnati 39° 5′ 54″ N. How many degrees farther south is Charleston than Cincinnati?

116. The longitude of Boston is 71° 4′ 20″; that of the city of Mexico 99° 5′. How many degrees farther west than Boston?

117. The longitude of Chicago is 87° 30′ 30″. How many degrees farther west is Chicago than Boston?

118. The longitude of St. Louis is 90° 15′ 10″. How many degrees farther west is St. Louis than Boston?

119. The latitude of Constantinople is 41° 0′ 16″ N. How many degrees farther south is Constantinople than Boston?

120. The latitude of St. Augustine is 29° 48′ 30″ N. How many degrees farther south is St. Augustine than Boston?

121. The latitude of Burlington, Vt., is 44° 27′; that of Washington, D. C., 38° 53′ 33″. How many degrees farther south is Washington than Burlington?

DIVISION OF DENOMINATE NUMBERS.

146. RULE. *Divide each denomination of the dividend, beginning at the left, by the divisor, and write the result in the quotient. Reduce the remainder, if there be any, to the next lower denomination, and add it to the number of the same denomination in the dividend. Continue the division, and write the result in the quotient, as before.*

PROOF. *The same as in division of simple numbers.*

OBS. 1. When the divisor exceeds 12, and is a composite number, divide by each factor in succession, as in simple division.

OBS. 2. Division of denominate numbers is precisely the same in principle as simple division. Reducing the remainder to the next lower denomination, and adding it to the next number, is the same as *prefixing* the remainder to the next figure.

122. Divide £266, 12 s. $9\frac{3}{4}$ d. into 6 equal parts.

	£.	s.	d.
6)	266	12	$9\frac{3}{4}$
	44	8	$9\frac{2}{4}$, 3 qr. rem.

6 in 26 four times and 2 over. Write the 4 in the quotient, and prefix the 2 to the next figure, making 26. 6 in 26 four times and 2 over. Write the 4 in the quotient, and reduce the 2 pounds to shillings, which are equal to 40 shillings, and add them to the next number, 12 shillings, making 52 shillings. 6 in 52 eight times and 4 remainder, which reduced to pence, and added to the 9, make 57 pence. 6 in 57 nine times and 3 over, which reduced to farthings, and added to the 3, make 15 farthings. 6 in 15 two times and 3 over.

What is the rule for the division of denominate numbers?

EXAMPLES.

123. Divide £329, 17 s. 6 d. 2 qr. by 14.

124. Divide £735, 13 s. 7 d. 3 qr. by 25.

125. Divide £964, 2 s. 1 d. 1 qr. by 27.

126. Divide 324 cwt. 2 qr. 3 lb. by 93.

127. Divide 312 cwt. 1 qr. 21 lb. by 63.

128. Divide 67 tons, 13 cwt. 2 qr. by 97.

129. Divide 365 days, 5 h. 48 m. 51 s. by 12.

130. If 36 cwt. of cheese cost £80, 2 s., how much will 1 cwt. cost?

131. If a gentleman's income be £548 a year, how much is his income for a month?

132. If 19 parcels of tea contain 332 lb. 8 oz., what is the weight of one parcel?

133. If 9 cows eat 21 tons, 7 cwt. 2 qr. 12 lb. of hay in a year, how much will 1 cow eat in the same time?

134. If 42 acres produce 567 bu. 3 pk. of oats, how much will 1 acre produce?

135. If 1 lb. of silver be worth $15.50, what would be the weight of $500,000 in silver?

136. If 1 ounce of pure gold be worth $18.60, what would be the weight of 1 million dollars in gold?

137. A farmer puts 1350 bushels of apples into 600 barrels. How many does he put in each barrel?

138. How many spoons, weighing 18 dwt. each, can be made of 5 lb. 9 oz. 12 dwt. of silver?

139. How many bottles, containing 1 pt. 2 gi. each, will hold 32 gal. 2 qt. of cider?

140. At 1 s. 6 d. per gal., how many gallons of molasses can be bought for £35, 18 s.?

141. If 39 boxes of oranges cost £202, 12 s. 6 d., what was the price per box?

142. If a man travel 21 m. 5 fur. in a day, how many days will it take him to travel 207 m. 3 fur.?

PRACTICAL QUESTIONS.

143. How much would 16 boxes of sugar cost, at $6\frac{1}{4}$ cents per lb., if each box contain 4 cwt. 3 qr. 18 lb.?

144. If a pile of wood be 140 ft. long, 3 ft. 6 in. wide, how high must it be to contain 18 cords?

145. If a gentleman's annual income be £1000, and his daily expenses £1, 17 s. $3\frac{1}{2}$ d., how much does he save in 9 years?

146. If a gentleman receive £1, 9 s. 9 d. per week, what is his annual income?

147. If the distance between London and Edinburgh be 389 m. 6 fur. 20 rods, how long would a person be in walking from one place to the other, at the rate of 27 m. 6 fur. 30 rods per day?

148. How many yards of cloth are there in 21 pieces, each piece containing 15 yd. 3 qr. 2 na.?

149. What will 25 doz. of knives and forks cost, at 11 s. 6 d. per doz.?

150. If 36 doz. of knives and forks cost £9, 10 s. 9 d., what costs 1 doz.?

151. If 45 yds. of superfine broadcloth cost £43, 10 s., how much is that a yard?

152. If 27 pairs of silk hose cost £8, 0 s. $10\frac{1}{2}$ d., how much is that per pair?

153. If 11 cwt. of sugar cost £22, 14 s. 8 d., what will 1 cwt. cost?

154. How long will a person be in saving £150, if he save 2 s. 6 d. per week?

155. If 16 acres produce 1246 bu. 3 pk. 6 qt. of corn, how much will 1 acre produce?

156. If the circumference of a wheel be 8 ft. 3 in., how many times will it turn in a distance of 184 miles?

157. How many dozen of tea-spoons, each spoon weighing 1 oz. 3 dwt., can be made out of 25 lb. 10 oz. 10 dwt. of silver?

147. To find the difference of time between two dates, —

RULE. *Write the first date under the last, the years on the left, and the number of the month in order next, and the day of the month on the right and then subtract.*

OBS. January is reckoned the first month, February the second, March the third, &c.

The difference of time between February 3, 1845, and September 19, 1846, is 1 year, 7 months, 16 days.

y.	mo.	d
1846	9	19
1845	2	3
1	7	16

OBS. In finding the difference between two dates, each month is usually reckoned 30 days.

158. What is the difference of time between September 4, 1845, and July 6, 1848?
159. What is the difference of time between October 9, 1844, and August 1, 1812?
160. What is the difference of time between November 29, 1842, and April 4, 1846?
161. What is the difference of time between May 18, 1845, and December 2, 1847?
162. What is the difference between June 6, 1845, and February 4, 1846?
163. What is the difference between January 1, 1840, and December 31, 1844?
164. What is the difference between August 1, 1842, and November 16, 1846?

148. To find the difference in time and longitude between different places, —

What is the rule for finding the time between two dates?

As the earth passes through 360° in 24 hours, in 1 hour it passes through 15°, and in 1 minute it passes through $\frac{1}{60}$ of 15°, or $\frac{1}{4}$ of a degree, or 15 geographical miles; and in one second it will pass through $\frac{1}{4}$ of a geographical mile. Therefore, *by multiplying the difference of longitude between two places, expressed in degrees, and minutes, and seconds, by 4, will give the difference of time in minutes, and seconds, and parts of a second.* Thus, if the difference of longitude between two places be 77°, the difference of time will be $77 \times 4 = 308$ minutes, equal to 5 hours, 8 minutes. If the difference of longitude be 24° 12′ 20″, the difference of time will be 24° 12′ 20″$\times 4 = 96$ minutes, 49 seconds, and $\frac{20}{60}$, or $\frac{1}{3}$, of a second, equal to 1 hour, 36 minutes, $49\frac{1}{3}$ seconds.

OBS. It is obvious that when it is noon at any particular place, at any point east of that place it is after noon, and at any point west of that place it is before noon.

165. If the difference of longitude between Boston and London be 70° 58′ 35″, what time is it in London when it is noon in Boston, and what time is it in Boston when it is noon in London?

166. If the longitude of Boston be 71° 4′ 20″, and that of Chicago 78° 30′ 30″, what is the difference in the time?

167. What time is it in Boston when it is noon at Chicago?

168. What time is it in Chicago when it is noon in Boston?

169. If a gentleman travel from Boston to Louisville, and his watch keeps accurately the Boston time, will his watch be too fast or too slow, on arriving at Louisville, and how much, allowing the difference of longitude to be 14° 25′?

What is the rule for finding the difference of time between wo places?

170. When it is noon at London, what time will it be at the mouth of Columbia River, which is 120° west of London?

149. When the difference of time between two places is known, the difference of longitude may be found *by dividing the minutes and seconds by* 4; *the quotient will be the difference of longitude in degrees and minutes.*

171. What is the difference of longitude between two places, if the difference of time be 5 h. 20 m. 16 sec.?

172. What is the difference of longitude between two places, if the difference of time be 2 h. 24 m. 15 sec.?

173. What is the difference of longitude between two places, if the difference of time be 7 h. 40 m. 20 sec.?

174. If a vessel sail from Boston for Europe, and, after a number of days, the captain finds, by taking an observation of the sun, that the difference of time, compared with his chronometer, which gives Boston time, is 2 h. 40 m., how many degrees is he east of Boston?

150. To reduce denominate numbers to equivalent decimals of a higher denomination, —

RULE. *Divide the lowest denomination by that number which makes one of the next higher denomination, and annex the quotient to the next higher denomination, and divide as before. Proceed thus through all the denominations to the last.*

OBS. 1. The denominate numbers may first be reduced to a common fraction, (ART. 158,) and then reduced to a decimal.

What is the rule for finding the difference of longitude? What is the rule for reducing denominate numbers to equivalent decimals of a higher denomination?

OBS. 2. The numbers should be written one above the other, the lowest denomination at the top and the highest at the bottom, and ciphers annexed when necessary.

175. Reduce £15 10 s. 9 d. to the decimal of a £.

```
12 )  9.00
20 ) 10.7500
     15.5375
```

The 9 pence is divided by 12, because 12 pence make 1 shilling. The quotient, 75, is annexed to the 10 shillings. This is divided by 20, because 20 shillings make £1.

EXAMPLES.

176. Reduce 16 s. 9¾ d. to the decimal of a pound.

177. Reduce 17 s. 5¾ d. to the decimal of a pound.

178. Reduce 18 s. 7½ d. to the decimal of a pound.

179. Reduce £58 12 s. 6½ d. to the decimal of a pound.

180. Reduce 2 oz. 14 dwt. 12 gr. to the decimal of a pound.

181. Reduce 2 qr. 17 lb. to the decimal of a hundred weight.

182. Reduce 1 R. 10 rds. to the decimal of an acre.

183. Reduce 365 d. 5 h. 48 m. 51 sec. to the decimal of a day.

184. Reduce 3 qr. 3 n. to the decimal of a yard.

185. Reduce 20 fur. 4 yds. to the decimal of a mile.

186. Reduce 26 sq. rd. to the decimal of an acre.

187. Reduce 3 pks. 7 qts. to the decimal of a bushel.

188. Reduce 20 min. 35 sec. to the decimal of a degree.

189. Reduce 4 h. 25 min. 34 sec. to the decimal of a day.

151. *Shillings*, *pence*, and *farthings* may also be reduced to a decimal of a pound in the following manner: *Write half the number of shillings in the place* *of tenths, and reduce the given pence and farthings to farthings, and if they amount to 24 or more, increase them by adding 1, and if there is an odd shilling, this sum is to be increased by 50, which must be written in the place of hundredths and thousandths.* Thus 12 s. 9 d., reduced to the decimal of a pound, is .637. The half of 12 is 6, and 9 d., changed to farthings, gives 36, which being increased by 1, because it exceeds 24, is made 37. Thus written, the places of hundredths and thousandths give .637.

The reason of the above rule is obvious from the fact, that as there are 20 shillings in a pound, half of the number of shillings will express the tenths; and as 1 farthing is $\frac{1}{960}$ of a pound, and 960, increased by $\frac{1}{24}$ of itself, is 1000, any number of farthings, increased by $\frac{1}{24}$ part of itself, will express so many thousandth parts of a pound. When the number of shillings is odd, 50 must also be added to the farthings, as 1 shilling is $\frac{50}{1000}$ of a pound.

Obs. When great accuracy is required, as many twenty-fourths of the farthings should be added to the farthings as there are farthings.

7 s. 6 d. = .375	18 s. 4 d. = .916	14 s. 3 d. = .712
3 s. 4 d. = .166	17 s. 9 d. = .887	12 s. 7 d. = .629
4 s. 6 d. = .225	15 s. 6 d. = .775	19 s. 1 d. = .954

152. To change decimals to denominate numbers, —

Rule. *Multiply the decimal by that number which is required of the next lower denomination to make a unit of the higher, and point off as in decimal fractions. Proceed thus with the decimal in each product. The figures on the left of the decimal point in the several products will be the denominate numbers.*

Obs. If there be a decimal in the last product, it should be changed to a common fraction, which will denote a fractional part of the lowest denominate number.

What is the rule for changing decimals to denominate numbers?

190. Reduce £.840625 to denominate numbers.

```
   .840625
        20
  ----------
 16.812500 £.
        12
  ----------
  9.750000 s.
         4
  ----------
  3.000000 d.
```

Multiply the decimal by 20, because 20 shillings make 1 pound. Point off from the right of the product six figures, as in the multiplication of decimals. Multiply the next decimal product by 12, because 12 pence make 1 shilling, and point off as before. Multiply the last decimal product by 4, because 4 farthings make 1 penny. The figures in the several products, at the left of the decimal point, are the denominate numbers, viz.: £16 9 s. 3 d.

EXAMPLES.

191. Reduce £.78125 to denominate numbers.

192. Reduce £.61925 to denominate numbers.

193. Reduce £.15625 to denominate numbers.

194. Reduce .728125 of a pound Troy to denominate numbers.

195. Reduce .9642857 of a month to denominate numbers.

196. What is the value of .45 of a bushel?

197. What is the value of .4 of a yard?

198. What is the value of .96 of a cord?

199. What is the value of .864 of a rod?

200. What is the value of .765 of a mile?

201. What is the value of .965 of a ton?

202. What is the value of .875 of an hour?

DENOMINATE FRACTIONS.

SECTION X.

153. Denominate fractions may be changed from a higher denomination to a lower, and from a lower to a higher, in the same manner as whole numbers.

154. To reduce denominate fractions from a higher denomination to a lower, —

Rule. *Multiply the fraction by the same numbers that are required to reduce a whole number of the same denomination as the fraction, to the lower denomination required*, (Art. 141.)

Obs. The numbers used as multipliers may first be changed to the form of an improper fraction, and factors common to the numerators and denominators may be cancelled.

1. Reduce £$\frac{1}{320}$ to the fraction of a penny.

$$\frac{1}{\underset{\underset{4}{\cancel{16}}}{\cancel{320}}} \times \frac{\cancel{20}}{1} \times \frac{\overset{3}{\cancel{12}}}{1} = \frac{3}{4}$$

The fraction is multiplied by 20 and 12, the same that are required in whole numbers, and the common factors are cancelled.

EXAMPLES.

2. Reduce £$\frac{1}{1536}$ to the fraction of a farthing.
3. Reduce £$\frac{1}{30}$ to the fraction of a shilling.
4. Reduce $\frac{1}{14400}$ of a lb. Troy to the fraction of a grain.

What is the rule for reducing denominate fractions from a higher denomination to a lower?

5. Reduce $\frac{3}{1280}$ of a lb. Avoirdupois to the fraction of a dram.
6. Reduce $\frac{11}{1344}$ of a cwt. to the fraction of a lb.
7. Reduce $\frac{1}{5824}$ of a ton to the fraction of a lb.
8. Reduce $\frac{3}{800}$ of an acre to the fraction of a rod.

155. To change denominate fractions from a lower denomination to a higher, —

RULE. *Divide the fraction by the same numbers that are required to change a whole number of the same denomination as the fraction, to the higher denomination required.*

OBS. The numbers used as divisors may be changed to improper fractions, and factors common to the numerators and denominators may be cancelled.

9. Reduce $\frac{5}{8}$ of a farthing to the fraction of a £.

$$\frac{\not{5}}{8} \times \frac{1}{4} \times \frac{1}{12} \times \frac{1}{\underset{4}{\not{2}\not{0}}} = \frac{1}{1536}$$

Dividing $\frac{5}{8}$ of a farthing by 4, 12, and 20, the same numbers that are required to change farthings to pounds, 5 is cancelled in the numerator, and as a factor of 20 in the denominators.

EXAMPLES.

10. Reduce $\frac{3}{4}$ of a penny to the fraction of a £.
11. Reduce $\frac{8}{9}$ of a farthing to the fraction of a £.
12. Reduce $\frac{2}{5}$ of a grain to the fraction of a lb. Troy.
13. Reduce $\frac{3}{5}$ of a dram to the fraction of a cwt.
14. Reduce $\frac{4}{5}$ of a nail to the fraction of a yard.
15. Reduce $\frac{8}{9}$ of a minute to the fraction of a day.
16. Reduce $\frac{4}{7}$ of a foot to the fraction of a mile.

What is the rule for changing denominate fractions from a lower denomination to a higher?

156. To reduce a fraction of any denomination to denominate numbers, —

RULE. *Multiply the numerator by the number required of the next lower denomination to make a unit of the denomination of the fraction, and divide the product by the denominator. If there be a remainder, multiply and divide in the same manner to the lowest denomination. The several quotients will be the denominate numbers required.*

17. What is the value of £$\frac{3}{7}$?

```
       3
      20
  ---------
  7 ) 60 ( 8 s.
      56
      --
       4
      12
  ---------
  7 ) 48 ( 6 d.
      42
      --
       6
       4
  ----------
  7 ) 24 ( 3 3/7 qr.
      21
      --
       3
```

Multiplying the numerator, 3, by 20, and dividing by the denominator, 7, gives 8 s. and 4 remainder. Multiplying the remainder by 12, and dividing again by 7, gives 6 d. and 6 remainder. Multiplying the remainder, 6, by 4, and dividing again by 7, gives 3 qr. and 3 remainder, which is $\frac{3}{7}$.

What is the rule for reducing a fraction of any denomination to denominate numbers?

EXAMPLES.

18. What is the value of £$\frac{5}{8}$?
19. What is the value of $\frac{11}{12}$ of a cwt.?
20. What is the value of $\frac{3}{9}$ of a lb. Troy?
21. What is the value of $\frac{7}{10}$ of a yard?
22. What is the value of $\frac{9}{11}$ of a mile?
23. What is the value of $\frac{2}{3}$ of a ton?
24. What is the value of $\frac{9}{11}$ of a year?
25. What is the value of $\frac{3}{5}$ of a cord?
26. What is the value of $\frac{1}{3}$ of $\frac{1}{4}$ of a bushel?
27. What is the value of $\frac{3}{4}$ of $\frac{5}{6}$ of a mile?
28. What is the value of .825 of a cwt.?
29. What is the value of .475 of a ton?
30. What is the value of .625 of a lb. Troy?

157. To reduce denominate numbers to equivalent common fractions,—

RULE. *Reduce the denominate number to the lowest denomination it contains, for a numerator, and a unit of the denomination of the required fraction to the same denomination as the numerator, for a denominator.*

31. Reduce 6 s. 9 d. 3 qr. to the fraction of a £.

6. 9. 3.	£1=20
12	12
81	240
4	4
327	960
960	

6 s. 9 d. 3 qr. reduced to farthings are 327 farthings. £1 reduced to the same denomination is 960. $\frac{327}{960}$ is the fraction required.

What is the rule for reducing denominate numbers to equivalent common fractions?

EXAMPLES.

32. Reduce 9 s. 11 d. 2 qr. to the fraction of a £.
33. Reduce 6 oz. 10 dwt. to the fraction of a lb.
34. Reduce 14 dwt. 12 gr. to the fraction of an oz.
35. Reduce 14 lb. 6 oz. 10 dr. to the fraction of an cwt.
36. What part of a ton is 4 qr. 16 lb. 8 oz.?
37. What part of a mile is $5\frac{1}{2}$ yd. and 9 ft?
38. What part of a mile is 4 fur. 30 rd. 16 ft.?
39. What part of a day is 4 h. 26 min.?
40. What part of a gallon is 2 qt. 3 gi.?
41. What part of a day is $\frac{3}{5}$ of an hour?
42. What part of a mile is $\frac{2}{3}$ of a rod?
43. What part of a yard is $\frac{3}{7}$ of a nail?

Obs. When the lowest denomination is expressed in the form of a fraction, the unit of the higher denomination must be reduced to such parts of the lowest denomination as are expressed by the denominator of the fraction.

158. To change one denominate number to the fractional part of any other denominate number, —

Rule. *Reduce the given numbers to the same denomination, and write the number which is to be the fractional part for the numerator, and the other number for the denominator of the fraction, and reduce the fraction to its lowest terms.*

44. What part of a £ is 16 s. 8 d.?

16 s. 8 d. £1×20×12=240
12

$$\frac{200}{240}=\frac{5}{6}$$

Reduce 16 s. 8 d. and £1 to pence. The fraction

What is the rule for reducing one denominate number to the fractional part of any other denominate number?

will be $\frac{200}{240}$, which reduced to its lowest terms will give $\frac{5}{6}$.

EXAMPLES.

45. What part of a £ is 9 s. 6 d.?
46. What part of a lb. Troy is 4 oz. 16 dwt.?
47. What part of a cwt. is 12 lb. 12 oz.?
48. What part of £4 is 10 s. 9 d.?
49. What part of 2 ton is 4 cwt. 3 qr. 20 lb.?
50. What part of 4 bushels is $\frac{1}{2}$ pt.?
51. What part of a mile is $\frac{1}{3}$ of a foot?
52. What part of 7 hours is $\frac{5}{6}$ of a minute?
53. What part of 3 gallons is $\frac{1}{3}$ of a gill?
54. What part of 2 cord is $10\frac{1}{2}$ cubic feet?
55. What part of 2 furlongs is $15\frac{1}{2}$ feet?
56. What part of 3 miles is $4\frac{1}{2}$ inches?

ADDITION AND SUBTRACTION OF DENOMINATE FRACTIONS.

159. Rule. *Reduce each fraction to equivalent denominate numbers, and add or subtract, as in* Arts. 143, 145.

Or reduce each fraction to the same denomination, and add or subtract, as in common fractions.

57. Add £$\frac{2}{3}$ and $\frac{2}{5}$s.

$$\begin{array}{rrrr} £\frac{2}{3} = & 13\text{s.} & 4\text{d.} & \\ \frac{2}{5}\text{s.} = & & 4\text{d.} & 3\frac{1}{5}\text{qr.} \\ \hline & 13\text{s.} & 8\text{d.} & 3\frac{1}{5}\text{qr.} \end{array}$$

$$\frac{2}{5} \times \frac{1}{20} \text{ or } \frac{1}{50} + \frac{2}{3} = \frac{103}{150} = 13\text{s. } 8\text{d. } 3\tfrac{1}{5}\text{qr.}$$

EXAMPLES.

58. Add £$\frac{4}{7}$, $\frac{1}{6}$ s., and $\frac{3}{5}$ d.
59. Add £$\frac{8}{9}$, $\frac{4}{7}$ s., and $\frac{4}{5}$ d.
60. Add $\frac{5}{6}$ lb. Troy and $\frac{4}{7}$ dwt.

What is the rule for adding and subtracting denominate fractions?

61. Add $\frac{1}{5}$ ton and $\frac{8}{9}$ lb.
62. Add $\frac{7}{9}$ gal. and $\frac{2}{3}$ pt.
63. Add $\frac{1}{3}$ m. and $16\frac{1}{2}$ ft.
64. Add $\frac{2}{7}$ ac. and $\frac{3}{7}$ rd.
65. From £$\frac{7}{9}$ take $\frac{4}{5}$ s.
66. From $\frac{5}{6}$ lb. Troy take $\frac{3}{7}$ oz.
67. From $\frac{1}{16}$ ton take $\frac{2}{3}$ lb.
68. From $\frac{1}{2}$ ac. take $\frac{4}{5}$ rd.
69. From $\frac{3}{4}$ m. take $\frac{9}{10}$ rd.
70. From $\frac{7}{8}$ yd. take $\frac{2}{3}$ na.

Obs. Denominate fractions may also first be reduced to decimals of the same denomination, and then added and subtracted as decimals.

PRACTICAL QUESTIONS.

71. What is the sum of £$\frac{3}{5}$ and 14 s. 9 d. 3 qr.

72. What is the sum of $\frac{4}{7}$ lb. Troy and 4 oz. 12 dwt.?

73. What is the sum of 4 tons, 2 qr. 12 lb. and 3.0245 tons?

74. What is the difference between 25 gal. 3 qt. and 14.0246 gal.?

75. If 4 cwt. 14 lb. of sugar cost \$30.50, what cost 1 lb.?

76. If .375 of a yard of cloth cost £$\frac{1}{3}$, what will 225 yards cost?

77. If 24 yd. 2 qr. cost \$125.50, what cost 1 yard?

78. If $67\frac{1}{2}$ gallons of molasses cost \$22.50, what cost 1 qt.?

79. If 1 gallon of molasses can be bought for £$\frac{1}{9}$, how many gallons can be bought for £20?

80. If 1 bushel of wheat cost 6 s. 9 d., how many bushels can be bought for £15, 16 s.?

81. How many square feet in a board that is 9 ft. 4 in. long and 1 ft. 6 in. wide?

82. Bought $16\frac{1}{2}$ bushels of salt at 75 cents per bushel, and sold it at 23 cents per peck. What was the gain?

83. A grocer bought a pipe of molasses, containing 124 gallons, at $33\frac{1}{3}$ cents per gallon; 19 gallons having leaked out, he sold the remainder at 40 cents per gallon. Did he gain or lose, and how much?

84. A farmer bought $28\frac{1}{3}$ cords of wood, at \$3.50 per cord; he sold 13 cords at \$4.30 per cord, and the remainder at \$3.20 per cord. Did he gain or lose, and how much?

85. Bought $\frac{3}{4}$ of an acre of land for \$325.50, and sold it at $\frac{2}{5}$ cents per foot? What was the gain or loss?

86. How many yards of carpeting, $\frac{3}{4}$ of a yard wide, will be required to carpet a floor that is 23 ft. 9 in. long, 16 ft. 8 in. wide?

87. How many square feet of paper will be required to cover the walls of a room that is 16 ft. 4 in. long, 15 ft. 6 in. wide, and 10 ft. 3 in. high, after deducting 116 ft. for windows and doors?

88. If the cargo of a ship be worth £8000, and if $\frac{3}{7}$ of $\frac{4}{5}$ of $\frac{7}{8}$ of the ship be worth $\frac{2}{3}$ of $\frac{3}{4}$ of $\frac{5}{8}$ of the cargo, what is the value of both ship and cargo?

89. A grocer bought 15 cwt. 3 qr. 21 lb. of coffee at \$9.50 per cwt., and sold it at $12\frac{1}{2}$ cents per lb. What did he gain on the whole?

90. A farmer purchased 4 acres of woodland, at \$150 per acre; he paid for cutting 300 cords of wood $\$\frac{3}{4}$ per cord; for carting the same, $\$1\frac{1}{2}$ per cord; he sold the 300 cords of wood at $\$4\frac{1}{2}$ per cord. Did he gain or lose by the bargain, and how much?

91. A farmer paid \$160.50 for the lease of a farm for one year; he sold 16 tons of hay, at \$13 per ton; 150 bushels of potatoes, at $33\frac{1}{3}$ cents per bushel; 156 bushels of corn, at 75 cents per bushel; 62 bushels of oats, at 42 cents per bushel; $20\frac{2}{3}$ barrels of apples, at \$1.75 per barrel; and $150\frac{4}{5}$ lb. of cheese, at 9 cents per pound. He paid for expenses of family and for labor \$345. Did he gain or lose, and how much?

PERCENTAGE.

SECTION XI.

160. Percentage, or Per Cent., denotes any number of hundredths of a given sum.

Obs. These terms are from two Latin words, *per* and *centum*, which signify *by the hundred*.

161. The per cent. is sometimes called *the rate* and is expressed in decimals, thus: —

1 per cent.	is	written	.01
2 per cent.	"	"	.02
3 per cent.	"	"	.03
4 per cent.	"	"	.04
5 per cent.	"	"	.05
6 per cent.	"	"	.06
7 per cent.	"	"	.07
8 per cent.	"	"	.08, &c.
$\frac{1}{2}$ of 1 per cent.	"	"	.005
$\frac{1}{4}$ of 1 per cent.	"	"	.0025
$\frac{1}{8}$ of 1 per cent.	"	"	.00125

Obs. When the parts of 1 per cent. cannot be expressed exactly in decimals, they should be written in the form of a common fraction, thus: $4\frac{1}{3}$ per cent. $=.04\frac{1}{3}$.

162. To find the per cent. of any number, —

Rule. *Multiply the given number by the per cent., and point off the product, as in multiplication of decimals.*

The principle of this rule is the same as that in the multiplication of decimals.

What does percentage, or per cent., denote? What are the terms derived from? What is the per cent. sometimes called, and how is it expressed? How are the parts of 1 per cent. written?

1. What is 4⅓ per cent. of 144?

```
 144
 .04⅓
 ----
 576
  48
 ----
6.24
```

Multiplying 144 by 4⅓, and pointing off according to the rule in the multiplication of decimals, gives 6.24.

EXAMPLES.

2. What is 6 per cent. of $34.556?
3. What is 5 per cent. of $90.755?
4. What is 4 per cent. of $875.25?
5. What is 3 per cent. of $456.675?
6. What is 2 per cent. of $640.205?
7. What is 1 per cent. of $735.75?
8. What is 10 per cent. of $905.60?
9. What is 6½ per cent. of $45.305?
10. What is 4⅓ per cent. of $734.02?
11. What is 1½ per cent. of $560.24?
12. What is ½ per cent. of $84.50?
13. What is ⅓ per cent. of $93.60?
14. What is ⅐ per cent. of $56.49?
15. What is ⅑ per cent. of $81.90?
16. A man paid 6½ per cent. for the use of $175.50 for 1 year. How much did he pay?
17. A man bought 15 shares of railroad stock, at $100 a share, and sold them at 9 per cent. advance? What did he gain?
18. A merchant bought 450 barrels of flour, at $4.37½ per barrel, and gained 12 per cent. in the sale of it. What did he receive?
19. A man's income is $1500 a year; he saves 12½ per cent. of it. How much does he spend?

163. To find what per cent. any number is of another given number,—

RULE. *Annex two ciphers to the number whose per cent. is sought, and divide by the number of which it is sought.* ART. 21. OBS.

OBS. The numbers must have the same number of decimal places, or be of the same denomination.

20. What per cent. of $50 is $3?

```
50 ) 300 ( 6
     300
     ---
```

Annexing two ciphers to the $3, and dividing by $50, gives 6 as the per cent.

OBS. This rule is evidently the converse of the preceding. Thus 6 per cent. of $50 is $3. 50 × .06 = 3.00.

21. What per cent. of $150 is $4?
22. What per cent. of $240 is $6?
23. What per cent. of $48 is $30?
24. What per cent. of $6250 is $.3125?
25. What per cent. of $840 is $648?
26. What per cent. of $56 is $.08?
27. What per cent. of $96 is $84?
28. What per cent. of $85 is $.065?
29. What per cent. of $506 is $.084?
30. What per cent. of $364.40 is $9?
31. What per cent. of $896.50 is $24?
32. What per cent. of $560 is $76.20?
33. What per cent. of $240 is $46.10?
34. What per cent. of $360 is $32.40?
35. What per cent. of $60.50 is $18?
36. What per cent. of £9 10 s. is £2 4 s. 9 d.?
37. What per cent. of £40 6 s. 9 d. is £5 10 s. 6 d.?

What is the rule for finding what per cent. any number is of another given number?

SIMPLE INTEREST.

SECTION XII.

164. INTEREST is a certain per cent. of a sum of money, paid for its use.

OBS. The per cent. is called the *rate*, and is always reckoned per annum, which signifies *by the year*.

165. The sum on which the interest is computed is called the *principal*. The principal and interest, added together, is called the *amount*.

166. The rate per cent. is established by law, and varies in different states and countries.

167. The legal rate is 6 per cent. in the New England States, also in New Jersey, Pennsylvania, Delaware, Maryland, Virginia, North Carolina, Tennessee, Kentucky, Ohio, Indiana, Illinois, Missouri, Arkansas, District of Columbia, and on debts and judgments in favor of the United States; also in Canada, Nova Scotia, and Ireland.

168. The legal rate is 7 per cent. in New York, Michigan, Wisconsin, Iowa, and South Carolina.

169. It is 8 per cent. in Georgia, Alabama, Mississippi, and Florida.

170. It is 5 per cent. in Louisiana, also in England and France.

171. RULE. *Find the interest of one dollar for the time, and by this multiply the principal;*

Or remove the decimal point in the principal two places to the left, and multiply by one half of the number of months, and one sixtieth of the number of days.

What is interest? What is the per cent. called? What is the meaning of per annum? What is the principal? What is the amount? How is the rate determined? What is the rule for casting interest?

OBS. If the principal contain only dollars, point off two figures on the right for cents.

It is evident that the interest of one dollar for twelve months, at 6 per cent., is *six cents;* for two months, or sixty days, *one cent;* and the interest of one dollar, for any number of months, is one cent. for every two months, or one half as many cents as there are months; and the interest of one dollar for one day is one sixtieth of a cent, or one sixth of a mill, and the interest of one dollar for any number of days is one sixth as many mills as there are days.

1. What is the interest of $1 for 1 year, 9 months, and 5 days?

Interest for 1 year	is =	.06
" " 9 months	is =	.045
" " 5 days	is =	$.000\frac{5}{6}$
		$.105\frac{5}{6}$

2. What is the interest of $1 for 3 years, 5 months, and 3 days?

Interest for 3 years	is =	.18
" " 5 months	is =	.025
" " 3 days	is =	$.000\frac{3}{6}$
		$.205\frac{3}{6}$

OBS. By the preceding rule, the interest for days is found for as many 360ths of one year's interest as there are days, which is evidently $\frac{5}{365}$, or $\frac{1}{73}$ too much. When great accuracy is required, as many 365ths of one year's interest must be taken as there are days, or $\frac{1}{73}$ of the interest for days, found by the preceding rule, must be deducted. Some states require this deduction to be made.

EXAMPLES.

3. What is the interest of $1 for 7 months and 9 days?

4. What is the interest of $1 for 10 months and 4 days?

5. What is the interest of $1 for 13 months and 1 day?

6. What is the interest of $1 for 17 months and 29 days?

7. What is the interest of $1 for 19 months and 2 days?

8. What is the interest of $645.60 for 2 years, 4 months, and 5 days?

645.60	2 years = 24 months.
$.140\frac{5}{6}$	4
2582400	2) 28
6456	.14
53800	$0\frac{5}{6}$
90.92200	$.140\frac{5}{6}$

In 2 years and 4 months there are 28 months. The interest of 1 dollar for 28 months is 14 cents; the interest for 5 days, $\frac{5}{6}$ of a mill, which added to 14 cents, gives $.140\frac{5}{6}$, the principal being multiplied by which gives $90.922.

Obs. When there are even months, and the days are less than 6, a cipher must be written in the place of mills.

9. What is the interest of $1557.56 for 3 years, 7 months, and 5 days?

10. What is the interest of $763.54 for 4 years, 9 months, and 11 days?

11. What is the interest of $351.67 for 3 years, 4 months, and 2 days?

12. What is the interest of $454.75 for 5 years, 5 months, and 11 days?

13. What is the interest of $570.60 for 6 years, 7 months, and 3 days?

14. What is the interest of $674.375 for 7 years, 3 months, and 27 days?

15. What is the interest of $735.75 for 8 years, 2 months, and 4 days?

16. What is the interest of $84.901 for 3 years, 1 month, and 6 days?

17. What is the interest of $936.40 for 4 years, 2 months, and 2 days?

18. What is the interest of $124.50 for 2 years, 4 months, and 5 days?

172. To find the amount of any sum of money for a given time, —

RULE. *Find the interest for the rate per cent. and time, and add it to the principal. Or multiply the principal by the amount of $1 for the time.*

19. What is the amount of $160 for 4 months and 12 days?

```
   160
  .022                    160
  ----                  1.022
   320                  -----
  320                     320
 -----                   320
 3.520                  160
 160                    --------
--------                $160.520
$163.520
```

Interest of $1 for 4 months and 12 days is 2 cents and 2 mills; the amount of $1 for the time is $1.022.

EXAMPLES.

20. What is the amount of $75.60 for 2 years, 11 months, and 1 day?

21. What is the amount of $324.60 for 3 years, 9 months, and 18 days?

22. What is the amount of $450.30 for 19 months and 29 days?

23. What is the amount of $675.80 for 1 year, 8 months, and 24 days?

What is the rule for finding the amount?

24. What is the amount of $735.75 for 2 years, 11 months, and 5 days?

25. What is the amount of $936.60 for 3 years, 4 months, and 2 days?

26. What is the interest of $84.54 for 7 years, 6 months, and 6 days?

27. What is the interest of $96.40 for 1 year, 9 months, and 5 days?

28. What is the amount of $244 for 4 years, 8 months, and 11 days?

29. What is the amount of $735.36 for 5 years, 2 months, and 17 days?

30. What is the interest of $560.25 for 2 years, 11 months, and 16 days?

31. What is the amount of $435.70 for 6 years, 10 months, and 14 days?

173. To find the interest for any sum of money, when the rate is any other than 6 per cent, —

RULE. *Find the interest at* 6 *per cent. for the given time, and divide the interest thus found by* 6, *which will give the interest for* 1 *per cent. Then multiply this quotient by the number denoting the per cent. sought.*

32. What is the interest of $344.40 for 3 years, 4 months, and 5 days, at 7 per cent.?

$$\begin{array}{r} \$344.40 \\ .200\frac{5}{6} \\ \hline 6888000 \\ 28700 \\ \hline 6\ )\ 6916700 \\ \hline 1152783 \\ 7 \\ \hline \$80.69481 \end{array}$$

What is the rule for finding the interest when the rate is any other than 6 per cent.?

3 years and 4 months are 40 months; interest for 40 months is 20 cents; interest for 5 days is $\frac{5}{6}$ of a mill; 20 cents plus $\frac{5}{6}$ of a mill is .200$\frac{5}{6}$, interest of $1 for the given time.

EXAMPLES.

33. What is the interest of $256.10 for 1 year, 9 months, and 3 days, at 7 per cent.?

34. What is the interest of $295.80 for 3 years, 7 months, and 5 days, at 8 per cent.?

35. What is the interest of $376.94 for 2 years, 3 months, and 2 days, at 5 per cent.?

36. What is the interest of $565.30 for 4 years, 5 months, and 4 days, at 4 per cent.?

37. What is the interest of $756.45 for 9 months and six days, at 3 per cent.?

38. What is the interest of $96.75 for 11 months and 29 days, at 2$\frac{1}{2}$ per cent.?

39. What is the amount of $739.40 for 2 years, 1 month, and 11 days, at 7 per cent.?

40. What is the interest of $84.20 for 3 years, 5 months, and 12 days, at 5 per cent.?

41. What is the amount of $96.30 for 5 years, 7 months, and 15 days, at 4$\frac{1}{2}$ per cent.?

42. What is the amount of $2452.06 for 7 months and 9 days, at 5$\frac{1}{4}$ per cent.?

43. What is the amount of $3764.08 for 11 months and 29 days, at 6$\frac{1}{2}$ per cent.?

44. What is the amount of $2460.90 for 93 days, at 7$\frac{1}{2}$ per cent.?

45. What is the amount of $643.73 for 63 days, at 3$\frac{1}{2}$ per cent.?

46. What is the amount of $960.40 for 25 days, at 7$\frac{1}{4}$ per cent.?

47. What is the amount of $65735 for 6 months and 3 days, at 8 per cent.?

48. What is the interest of $245.60 from July 2, 1848, to June 20, 1849?

49. What is the interest of $36.40 from August 10, 1847, to September 8, 1849?

50. What is the interest of $760.30 from October 8, 1846, to November 5, 1848?

51. What is the amount of $470.90 from May 6, 1844, to April 9, 1846?

52. What is the amount of $48.64 from March 1, 1846, to February 28, 1848?

53. What is the amount of $276 from January 1, 1848, to December 30, 1849?

54. What is the amount of $8650 from January 4, 1844, to September 6, 1847, at $5\frac{1}{2}$ per cent.?

55. What is the interest of $96.50 from February 7, 1843, to July 21, 1850, at $6\frac{1}{2}$ per cent.?

56. What is the interest of $84.60 from June 1, 1847, to May 29, 1849?

174. Interest may also be computed by the following rule: —

RULE. *Multiply the principal by the rate per cent., and the product will be the interest for 1 year; and multiply this product by the number of years for which the interest is required.*

For months, take such a fractional part of the interest for one year as is denoted by the number of months.

For days, take such a fractional part of the interest for one month as is denoted by the number of days.

OBS. Some accountants find the greatest number of whole months between the two dates, and the remaining time is reckoned in days, allowing as many days for each month as there are in each calendar month. Thus from January 15 to June 6 there are 4 months and 22 days. From January 15 to May 15 there are 4 months; allowing 31 days in May, there are 16 days remaining, which, added to the 6 in June, make 22 days.

What is the second rule for casting interest?

57. What is the amount of $764.20 for 4 years, 5 months, and 8 days?

	$764.20
	.06
Interest for 1 year,	458520
	.4
Interest for 4 years,	183.4080
Interest for 4 mos., ($\frac{1}{3}$ of 1 year,)	15.2840
Interest for 1 mo., ($\frac{1}{4}$ of 4 mos.,)	3.8210
Interest for 6 days, ($\frac{1}{5}$ of 1 mo.,)	7642
Interest for 2 days, ($\frac{1}{3}$ of 6 days,)	2547
	203.5319
	764.20
Amount,	$967.7319

Obs. 1. Great care must be used in pointing off the decimals, and in writing dollars under dollars, cents under cents, &c.

Obs. 2. When the per cent. is not specified, it is always considered to be 6 per cent.

EXAMPLES.

58. What is the interest of $456.84 for 7 years, 7 months, and 5 days?

59. What is the interest of $78.50 for 6 years, 9 months, and 12 days?

60. What is the interest of $96.60 for 5 years, 3 months, and 25 days?

61. What is the interest of $8735.69 for 11 years, 7 months, and 27 days?

175. Rule. *To find the interest on pounds, shillings, pence, and farthings, first reduce the shillings to the decimal of a pound, and proceed as in federal money.*

What is the rule for finding the interest on pounds, shillings, pence, and farthings?

EXAMPLES.

62. What is the interest of £25, 6 s. 11 d. for 2 years, 7 months, and 10 days?

63. What is the interest of £33, 11 s. $4\frac{3}{4}$ d. for 3 years, 7 months, and 18 days?

64. What is the amount of £121, 15 s. $9\frac{1}{2}$ d. for 4 years, 9 months, and 13 days?

65. What is the amount of £130, 19 s. $6\frac{1}{4}$ d. for 1 year, 11 months, and 25 days?

66. What is the interest on a note of six hundred and seventy-five dollars and fifty cents from May 4, 1844, to February 14, 1846?

67. What is the interest on a note of three hundred and forty-eight dollars and thirty cents from November 17, 1843, to September 12, 1846?

68. What is the interest on a note of fifty-seven dollars and sixty-four cents from November 16, 1843, to April 6, 1846, at 7 per cent.?

69. What is the amount due on a note of two hundred and forty-three dollars and twenty-nine cents from October 4, 1842, to March 18, 1845, at 5 per cent.?

70. What is the amount due on a note of four hundred and fifteen dollars and twenty cents from August 5, 1846, to January 24, 1849, at 7 per cent.?

71. What is the interest on a note of one hundred and eighteen dollars and twenty cents from February 28, 1844, to July 17, 1847, at 7 per cent.?

72. What is the interest on a note of seventy-six dollars and forty-four cents from June 12, 1845, to October 16, 1847?

73. What is the amount of a note of thirty-four dollars and ninety cents from October 4, 1843, to December 15, 1847?

74. What is the amount of a note of ninety-six dollars and forty cents from August 10, 1845, to July 15, 1848?

PARTIAL PAYMENTS.

176. The following rule has been adopted by the Supreme Court of the United States, and by many of the states, for computing interest on bonds and notes when partial payments have been endorsed on them: —

RULE. "*The rule for casting interest, when partial payments have been made, is to apply the payment, in the first place, to the discharge of the interest then due. If the payment exceeds the interest, the surplus goes towards discharging the principal, and the subsequent interest is to be computed on the balance of principal remaining due.*

"*If the payment be less than the interest, the surplus must not be taken to augment the principal; but interest continues on the former principal until the period when the payments, taken together, exceed the interest due, and then the surplus is to be applied towards discharging the principal, and interest is to be computed on the balance, as aforesaid.*"

$400. PROVIDENCE, *April* 12, 1840.

75. For value received, I promise to pay to James Fenner, or order, four hundred dollars, on demand, with interest.

ALFRED BLODGET.

On this note were received the following endorsements: —

March 4, 1842, received forty-five dollars.
May 10, 1843, received sixty dollars.
June 6, 1845, received ninety dollars.

What was due January 12, 1847?

What is the rule for casting interest when partial payments are made?

Principal,	$400
Interest to 1st payment, (22 m. 22 d.)	45.466
Payment being less than the interest, interest to the 2d payment, on same principal, (14 m. 6 d.)	28.400
Amount to 2d payment,	473.866
Sum of both payments, 45+60=	105.000
	368.866
Interest to the next payment, (24 m. 26 d.)	45.862
	414.728
Last payment,	90.000
	324.728
Interest to January 12, 1847, (18 m. 6 d.)	31.173
Balance remaining due January 12, 1847,	$355.901

$600.75. NEW YORK, *May* 4, 1840.

76. For value received, I promise to pay to James Cunningham, or order, six hundred dollars and seventy-five cents, on demand, with interest.

BENJAMIN TUCKER.

On this note were endorsed the following payments: —

Oct. 6, 1841, received sixty-four dollars.

July 8, 1842, received forty-eight dollars and fifty cents.

Nov. 20, 1844, received two hundred dollars and sixty cents

What was due May 10, 1846?

$565.90. BUFFALO, *June* 12, 1840.

77. For value received, I promise to pay William Bliss, or order, five hundred and sixty-five dollars and ninety cents, on demand, with interest.

GEORGE PACKARD.

On this note were endorsed the following payments: —

Jan. 6, 1841, received forty-five dollars.

March 6, 1843, received sixty-eight dollars.

Sept. 9, 1844, received thirty-seven dollars.

What was due December 30, 1846?

$340.00. PROVIDENCE, *July* 9, 1841.

78. For value received, I promise to pay to Henry Blackstone, or order, three hundred and forty dollars, on demand, with interest.

NOAH CURTIS.

On this note were endorsed the following payments: —

May 4, 1842, received sixty-six dollars and ninety cents.

August 10, 1843, received twenty-nine dollars and four cents.

Oct. 12, 1844, received forty-two dollars and six cents.

What was due November 20, 1845?

$609.65. PHILADELPHIA, *June* 8, 1845.

79. For value received, I promise to pay to Benjamin Tucker, or order, six hundred and nine dollars and sixty-five cents, in six months, with interest afterwards.

ARTEMAS SMITH.

On this note were endorsed the following payments: —

Oct. 4, 1846, received twenty-five dollars.

March 15, 1847, received sixteen dollars and twenty-five cents.

August 24, 1848, received thirty-six dollars and fifty-six cents.

What was due December 19, 1849?

$874.95. New York, *May* 9, 1843.

80. For value received, I promise to pay to Horatio Tremlet, or order, eight hundred and seventy-four dollars and ninety-five cents, in three months, with interest afterwards.

Thomas Cheever.

On this note were received the following endorsements: —

April 12, 1844, received fifty-six dollars and thirty cents.

July 14, 1845, received twenty-four dollars and eighty cents.

Sept. 18, 1846, received two hundred and forty dollars and sixty cents.

What was due February 10, 1848?

177. The following, though not a legal rule, is adopted by many for computing interest when the note on which partial payments have been made is settled within a year from the time the interest commenced: —

Rule. *Find the amount of the note for the whole time. Then find the amount of each payment, from the time it was endorsed, to the time of settlement. Subtract the amount of the several payments from the amount of the note.*

Obs. This rule is adopted in *Vermont* for any period of time.

$240. Providence, *May* 4, 1845.

81. For value received, I promise to pay to Joshua Bent, or order, on demand, two hundred and forty dollars, with interest.

Hiram Bradley.

What is the rule for finding the interest on notes settled within a year?

On this note were received the following endorsements : —

Sept. 10, 1845, received sixty dollars.
Jan. 16, 1846, received ninety dollars.

What was due May 4, 1846 ?

1st payment,	$60	2d payment,	$90	Principal,	$240
Int. 7m. 24d.,	2.34	Int. 3m. 18d.,	1.62	Int. 12m.,	14.40
Amount,	62.34		91.62		254.40
			62.34		153.96
Amount of payment,			153.96	Balance,	100.44

$460. CINCINNATI, *September* 10, 1846.

82. For value received, I promise to pay to Joseph Hovey, or order, on demand, four hundred and sixty dollars, with interest.

PHINEAS BROWN.

On this note were received the following endorsements : —

Jan. 4, 1847, received two hundred dollars.
May 15, 1847, received sixty-five dollars.

What is due September 10, 1847 ?

$340. PORTLAND, *June* 16, 1848.

83. Three months after date, I promise to pay to Jacob Appleton, or order, three hundred and forty dollars, with interest.

WILLIAM MORSE.

On this note were received the following endorsements : —

Oct. 14, 1848, received eighty-six dollars.
Feb. 12, 1849, received forty dollars.

What was due August 10, 1849?

$400. HARTFORD, *September* 12, 1848.

84. Four months after date, I promise to pay to Charles Newcomb, or order, four hundred dollars, with interest.

HENRY GAY.

On this note were received the following endorsements: —

Dec. 12, 1848, received one hundred and ten dollars.

March 16, 1849, received eighty-six dollars.

What is due October 9. 1849?

178. The following is the Connecticut rule: —

"*Compute the interest on the principal to the time of the first payment; if that be one year or more from the time the interest commenced, add it to the principal, and deduct the payment from the sum total. If there be after payments made, compute the interest on the balance due to the next payment, and then deduct the payment as above, and in like manner from one payment to another, till all the payments are absorbed, provided the time between one payment and another be one year or more.*

"*If any payments be made before one year's interest has accrued, then compute the interest on the principal sum due on the obligation for one year, add it to the principal, and compute the interest on the sum paid, from the time it was paid, up to the end of the year; add it to the sum paid, and deduct that sum from the principal and interest, added as above.*

"*If a year extends beyond the time of payment, then find the amount of the principal remaining unpaid up to the time of settlement, likewise the amount of the endorsements, from the time they were paid to*

What is the Connecticut rule?

the time of settlement, and deduct the sum of these several amounts from the amount of the principal. If any payments be made of less sum than the interest arisen at the time of such payment, no interest is to be computed, but only on the principal sum for any period."

Obs. 1. Cast the interest on all of the notes by each of the preceding rules.

Obs. 2. In some states when a note is written with "interest annually," interest is computed on the principal to the time of settlement, and on each year's interest after it is due, and the sum of the interests is added to the amount of the principal.

PROBLEMS IN INTEREST.

179. The interest, time, and rate per cent. being given, to find the principal, —

Rule. *Divide the interest by the product of the rate per cent. and time, and the quotient will be the principal.*

Obs. 1. It must be remembered that the rate per cent is a decimal, and expresses a certain number of hundredths. The quotient must therefore be pointed off according to the rule of division of decimal fractions.

Obs. 2. Months and days must be reduced to a decimal of a year, or to a fractional part of a year.

85. What principal will gain $78.40 in 2 years and 4 months, at 6 per cent.?

$$\frac{78.40}{2\frac{1}{3} \times .06} = 560.$$

Divide the interest, $78.40, by the product of the rate and time. The time, 2 years and 4 months, is $2\frac{1}{3}$ years, which, multiplied by .06, is .14. As 6 is a decimal denoting hundredths, the product of $2\frac{1}{3}$ by .06 will be 14 hundredths. 78.40 divided by .14 will be

What is the rule when the interest, time, and rate are given?

$560, since there are as many decimal places in the dividend as divisor.

EXAMPLES.

86. What sum must be invested to gain $450 at 6 per cent. in 9 months?

87. What principal will gain $750 at 6 per cent. in 1 year and 3 months?

88. What sum must be invested at 7 per cent. to gain $760 in 8 months?

89. What sum will pay a semi-annual dividend of $640 at 8 per cent.?

90. What sum invested at 7 per cent. will produce $900 a year?

180. The principal, interest, and rate per cent. being given, to find the time, —

RULE. *Divide the interest by the product of principal and rate, and the quotient will be the time.*

91. In what time will $640 gain $102.40 at 6 per cent.?

$$\frac{102.40}{640 \times .06} = 2\tfrac{2}{3} \text{ years} = 2 \text{ years and } 8 \text{ months.}$$

Dividing the interest, $102.40, by the product of the principal and rate, gives $2\frac{2}{3}$ years for the time. $640 \times .06 = 38.40$. Two figures for decimals, as 6 per cent. is 6 hundredths. As there are the same number of decimal places in the dividend and divisor, the quotient will be a whole number.

EXAMPLES.

92. In what time will $500 gain $500 at 6 per cent.?

93. In what time will $460 gain $230 at 6 per cent.?

What is the rule when the principal, interest, and rate are given?

94. In what time will $330 gain $110 at 6 per cent. ?

95. In what time will $68.50 gain $34.25 at 6 per cent. ?

96. How long will it take $600 to gain $600 at 5 per cent. ?

97. How long will it take $800 to gain $800 at 7 per cent. ?

98. How long will it take $400 to gain $400 at 10 per cent. ?

181. The principal, interest, and time being given, to find the rate per cent., —

RULE. *Divide the interest by the product of the principal and time, and the quotient will be the rate per cent.*

99. The interest of $400 for 3 years and 6 months is $84. What is the rate per cent. ?

$$\frac{84.00}{400 \times 3\frac{1}{2}} = .06, \text{ or 6 per cent.}$$

Divide the interest, $84, by the product of the time and principal. The product of 400 and $3\frac{1}{2}$ is 1400. Since 1400 is not contained in 84, two ciphers must be annexed for decimals. As the dividend contains two more decimal places than the divisor, there must be the same number pointed off in the quotient. The quotient is therefore 6 one hundredths, or 6 per cent.

EXAMPLES.

100. At what rate per cent. will $840 gain $49 in 2 years and 2 months ?

101. A man loaned $750 for 3 years and 4 months, and received $160.50 for the use of it. What did he receive per cent. ?

What is the rule when the principal, interest, and time are given?

102. A man deposited in a savings bank $84, for which he received $2.10 for 6 months. What per cent. did he receive?

103. If $500 be received as a semiannual dividend on an investment of $100,000, what per cent. is the dividend?

OBS. These rules may be concisely represented by the following formulas: —

Let p represent the principal, t the time, r the rate per cent., and the interest, as follows: —

p = principal.
t = time.
r = rate per cent.
i = interest.

The first rule may be represented by the formula $\frac{i}{t \times r} = p$.

The second rule by the formula $\frac{i}{p \times r} = t$.

The third rule by the formula $\frac{i}{p \times t} = r$.

COMPOUND INTEREST.

SECTION XIII.

182. COMPOUND INTEREST is the interest on the principal, and on the interest added to the principal after it becomes due.

RULE. *Find the interest on the principal to the time the interest becomes due, and add it to the principal. Then find the interest on this sum for the next period, and add the interest as before. Proceed in this manner with each successive period at which the interest becomes due. Subtract the first principal from the last sum, and the remainder will be the compound interest.*

What is the rule for compound interest?

OBS. When there are months and days, find the interest for them on the principal for the last period.

1. What is the compound interest of $300 for 3 years, at 6 per cent.?

	$300
Interest for one year,	18
Principal for the second year,	318
Interest for the second year,	19.08
Principal for the third year,	337.08
Interest for the third year,	20.2248
	357.3048
The first principal subtracted,	300
Compound interest for 3 years,	$57.3048

EXAMPLES.

2. What is the compound interest of $600 for 2 years, to be paid semiannually?

3. What is the compound interest of $320 for 2½ years, at 7 per cent.?

4. What is the compound interest of $540 for 3 years, at 5 per cent.?

5. What is the compound interest of $840 for 2 years, at 7 per cent., to be paid quarterly?

6. What is the compound interest of $460 for 3 years, 4 months, and 10 days?

183. The process of computing compound interest may be much shortened by the following table, in which the amount of $1 is computed for 30 years, at 4, 5, 6, and 7 per cent.

184. To find the amount of any sum by the table,—

RULE. *Multiply the given sum by the amount of* $1 *for the time, as found in the table.*

TABLE,

Showing the amount of 1 *dollar, or* 1 *pound, for any number of years under* 30, *at* 4, 5, 6, *and* 7 *per cent., compound interest.*

Years.	4 per cent.	5 per cent.	6 per cent.	7 per cent.
1	1.040000	1.050000	1.060000	1.070000
2	1.181600	1.102500	1.123600	1.144900
3	1.124864	1.157625	1.191016	1.225043
4	1.169858	1.215506	1.262476	1.310795
5	1.216652	1.276281	1.338225	1.402552
6	1.265319	1.340095	1.418519	1.500730
7	1.315931	1.407100	1.503630	1.605781
8	1.368569	1.477455	1.593848	1.718186
9	1.423311	1.551328	1.689478	1.838459
10	1.480284	1.628894	1.790847	1.967151
11	1.539454	1.710339	1.898298	2.104852
12	1.601032	1.795856	2.012196	2.252191
13	1.665073	1.885649	2.132928	2.409845
14	1.731676	1.979931	2.260903	2.578534
15	1.800943	2.078928	2.396558	2.759032
16	1.872981	2.182874	2.540351	2.952164
17	1.947900	2.292018	2.692772	3.158815
18	2.025816	2.406619	2.854339	3.379932
19	2.106849	2.526950	3.025599	3.616528
20	2.191123	2.653297	3.207135	3.869685
21	2.278768	2.785962	3.399563	4.140563
22	2.369918	2.925260	3.603537	4.430403
23	2.464715	3.071523	3.819749	4.740530
24	2.563304	3.225099	4.048934	5.072367
25	2.665836	3.386354	4.291870	5.427434
26	2.772469	3.555672	4.549382	5.807352
27	2.883368	3.733456	4.822345	6.213868
28	2.998703	3.920129	5.111686	6.648838
29	3.118651	4.116135	5.418387	7.114257
30	3.243397	4.321942	5.743491	7.612255

EXAMPLES.

7. What is the amount of $900 for 5 years, at 7 per cent.?

8. What is the amount of $640 for 3 years and 6 months, at 5 per cent.?

9. What is the amount of $10,000 for 30 years, at 6 per cent.?

10. What is the amount of $5000 for 20 years, at 5 per cent.?

DISCOUNT.

SECTION XIV.

185. Discount is a deduction of a certain per cent. from a sum of money, when paid before it is due.

186. The sum paid is called the *present worth*.

187. To find the present worth of any sum of money, and its discount for a given time, —

Rule. *Divide the sum by the amount of $1 for the rate and time, and the quotient will be the present worth. Subtract the present worth from the sum, and the remainder will be the discount.*

Obs. This rule may be expressed by the following formula: — The present worth (p) is equal to the given sum or amount (a) divided by 1 plus the product of the time (t) and the rate (r). Thus $p = \frac{a}{1+t\times r}$. $a - p =$ the discount.

The reason of this rule is obvious. Since $1 is the present worth of its amount, the present worth of any sum will be as many dollars as the amount of 1 dollar is contained in it.

What is the rule to find the present worth of any sum of money, and its discount for any given time?

1. What is the present worth of $336, payable in 10 months, when money is worth 6 per cent.?

The amount of $1 for 10 months is $1.05.
$336 ÷ 1.05 = $320, present worth.

EXAMPLES.

2. What is the present worth of $464, payable in 16 months, when money is worth 6 per cent.?

3. What is the present worth of $936, payable in 1 year and 10 months, when money is worth 7 per cent.?

4. What is the discount on $3200, payable in 9 months, at 6 per cent.?

5. What is the discount on $840, payable in 1 year and 3 months, at 6 per cent.?

6. A merchant bought a quantity of goods for $560.40 cash, and sold them the same day for $640.80 on 9 months' credit, when money was worth 6 per cent. How much did he gain on the goods?

7. What is the difference between the interest and the discount of $5000 for 1 year and 6 months?

188. It is now almost the universal practice of merchants and others to find the discount by the following rule: —

Rule. *Deduct the simple interest of the given sum for the rate and time, the difference will be the present worth.*

EXAMPLES.

8. What is the present worth of $90 for 2 years and 4 months?

9. What is the present worth of $175.60 for 3 years and 9 months, at 7 per cent.?

10. What is the present worth of $840 for 6 months and 10 days?

11. What is the present worth of $56.75 for 2 years, at 1 per cent. a month?

12. A merchant sold $1500 worth of goods, one half to be paid in 6 months, the other half in 9 months. What sum must he receive for them in cash, after deducting $1\frac{1}{2}$ per cent. a month?

BANK DISCOUNT.

189. *Bank discount* of a note, &c., is the interest of the sum specified in the note from the time the note is discounted to the time it becomes due, with three days additional, called *days of grace.*

190. To find the bank discount on a note, draft, &c., —

RULE. *Find the interest on the sum specified in the note for the time, including the three days of grace. Subtract the discount from the sum contained in the note, and the remainder will be the present worth.*

13. What is the present worth of $400 for 90 days, at 6 per cent.?

.0155	400.00
400	6.20
$6.2000, bank discount.	$393.80, present worth.

The interest of $1 for the 93 days is .0155. $400 multiplied by this interest gives the bank discount, $6.20. This, subtracted from $400, gives the present worth.

What is the rule to find the bank discount on a note, draft, &c.?

EXAMPLES.

14. What is the bank discount on a note for $350 of 60 days?

15. What is the bank discount on a note for $560 of 90 days?

16. What sum will be received from a bank for a note of $640 for 4 months?

17. What sum will be received from a bank for a note of $2500, payable in 6 months?

18. What is the bank discount on a note for $5000, payable in 9 months?

191. To find the sum for which a note must be given at a bank, in order to obtain a certain sum for a given time, —

Rule. *Divide the sum to be obtained by the present worth of* $1 *for the given rate and time, at bank discount, and the quotient will be the sum required.*

It is evident, since the present worth of $1 requires $1 to be discounted, as many dollars will be required to be discounted for any sum as the present worth of $1 is contained in this sum.

19. For what sum must a note be given, to be paid in 60 days, in order to obtain $460 from a bank, at 6 per cent.?

	$1.0000
Interest of $1 for 63 days,	.0105
Present worth of $1,	.9895

$460÷.9895=$464.881.

The interest of $1 for 63 days, subtracted from $1, gives the present worth of $1. $460, the sum required, divided by this, gives the sum for which the note must be given.

What is the rule to find the sum for which a note must be given at a bank to obtain a certain sum for a given time?

EXAMPLES.

20. For what sum must a note be written, in order to receive from a bank $500 for 6 months?

21. For what sum must a note be written, in order to receive from a bank $1000 for 9 months?

22. For what sum must a note be written, in order to receive from a bank $400 for 90 days?

23. A merchant bought a quantity of goods for $600. For what sum must he write his note, to be discounted at the bank for 6 months?

24. A farmer bought a farm for $5000 cash, and having only half of the sum on hand, he wishes to obtain the balance from the bank. For what sum must a note be written, to be discounted for 9 months?

COMMISSION.

SECTION XIV.

192. COMMISSION is the compensation paid to agents for their services in transacting business for others. It is usually estimated at a certain per cent. of the money employed in the transaction.

OBS. The agents are styled *factors*, *brokers*, *commission merchants*, *correspondents*, &c.

193. To find the commission on any sum of money, —

RULE. *Multiply the given sum by the per cent., and point off as in decimals.*

EXAMPLES.

1. What is the commission on $450.60, at 2½ per cent.?

2. What is the commission on $886.50, at 3½ per cent.?

What is commission? Recite the rule.

3. What is the commission on $630.40, at $4\frac{1}{2}$ per cent.?

4. A commission merchant sold goods amounting to $6560.75, at $1\frac{1}{2}$ per cent. What was his commission?

5. An auctioneer sold goods at auction, amounting to $376.40, at $2\frac{1}{2}$ per cent. What was his commission?

6. An auctioneer sold goods at auction, amounting to $640.50, at $2\frac{3}{4}$ per cent. What was his commission?

7. A collector was entitled to $4\frac{1}{2}$ per cent. on the money collected. How much commission did he receive on $2060?

8. A commission merchant sold goods amounting to $75,600, at $3\frac{1}{3}$ per cent. What was his commission?

194. To find the commission when it is to be deducted from the given sum, and the remainder to be invested, —

RULE. *Divide the given sum by $1 plus the given rate of commission, and the quotient will be the sum to be invested. Subtract the sum to be invested from the given sum, and the remainder will be the discount.*

OBS. This rule is the same as the rule for discount, with the exception of the time.

It is evident that the given sum must contain $1 plus the commission as many times as there are dollars to be invested.

EXAMPLES.

9. A gentleman sent to a broker $1218, to be invested in railroad stock, after deducting his commission of $1\frac{1}{2}$ per cent. How much did the broker receive for commission, and how much did he invest?

What is the rule to find the commission on a given sum, to be invested?

10. A merchant sent to his agent $2440, to purchase goods, after deducting his commission of $2\frac{1}{2}$ per cent. How much was his commission, and what sum did he spend for goods?

11. A merchant sent to his agent in Buffalo $5049, to purchase flour. How many barrels of flour did he purchase, at $$4\frac{1}{8}$ per barrel, after deducting his commission of 2 per cent.?

12. A broker negotiated a bill of exchange of $6015 for $\frac{1}{4}$ per cent. How much commission did he receive?

STOCKS.

SECTION XV.

195. THE money employed by companies in trade is styled *stock*, as bank stock, railroad stock, manufacturing stock, &c.; also government bonds and funds are styled stocks.

196. The whole amount invested by any company or corporation is called the *capital stock*, which is usually divided into *shares*.

197. The first cost of a share is called its *par value.*

OBS. *Par* is from a Latin word, signifying equal.

198. When stocks will sell for their first cost, they are said to be *at par.* When they will sell for more than their first cost, they are said to be *above par*, or *at a premium.* When they will not sell for their first cost, they are said to be *below par*, or *at a discount.*

199. The profits of stock are called *dividends*, which are paid at regular periods to the owners or stockholders.

What is stock? What is capital stock? What is par value? When are stocks said to be at par? When above par? When below par? What are the profits of stock called?

200. To find the value of stocks, —

Rule. *When the stocks are above par, multiply the sum invested by $1 plus the per cent., and the product will be their value.*

When the stocks are below par, multiply the sum invested by $1 minus the per cent., and the product will be their value.

EXAMPLES.

1. What is the value of $900 in stock, at 6 per cent. above par?
2. What is the value of $1000 in stock, at 3½ per cent. below par?
3. What is the value of $500 in stock, at 10½ per cent. below par?
4. What sum must be paid for 20 shares in the Western Railroad stock, at 5 per cent. above par, the par value of each share being $100?
5. What sum must be paid for 50 shares in the Worcester Railroad, at 3½ per cent. below par, the par value of each share being $100?

INSURANCE.

SECTION XVI.

201. Insurance is a security given to pay a certain sum on ships, houses, or property of any kind that may be destroyed by fire, accidents, or at sea.

202. The sum paid for the insurance is called the *premium*, and is estimated at a certain per cent. of the value of the property insured.

203. The written agreement or certificate is called the *policy.*

What is the rule to find the value of stocks? What is insurance? What is the premium? What is the policy?

204. The company or persons insuring are called *underwriters.*

Obs. Property is seldom insured for its whole value.

205. To find the premium on any amount of property insured, —

Rule. *Multiply the sum to be insured by the per cent., and the product will be the premium.*

This rule is the same in principle as percentage.

EXAMPLES.

1. What is the premium for insuring \$4000 on a house, at $1\frac{3}{4}$ per cent. ?
2. What premium must be paid for insuring \$600 on a barn, at $2\frac{7}{8}$ per cent. ?
3. What premium must be paid for insuring \$9000 on a store, at $1\frac{3}{8}$ per cent. ?
4. What premium must be paid for insuring a cargo of cotton from New Orleans to Liverpool, valued at \$9600, at 1 per cent. ?
5. What premium must be paid for insuring a cargo of sugar from Havana to St. Petersburg, valued at \$25,000, at $2\frac{1}{4}$ per cent. ?
6. If a vessel and cargo, valued at \$65,000, and insured at $4\frac{1}{4}$ per cent., were lost, what would be the loss to the underwriters ?
7. What sum must be insured on a vessel and cargo, valued at \$40,000, at $5\frac{1}{2}$ per cent, in order to include the premium and the property insured ?
8. What sum must be insured on \$70,000, to include the premium of $4\frac{1}{2}$ per cent. and a commission on the property insured of $\frac{3}{4}$ per cent. ?

What are underwriters ? Recite the rule to find the premium on property insured.

PROFIT AND LOSS.

SECTION XVII.

206. PROFIT AND LOSS treats of the gain and loss in trade, and shows how the price of goods must be adjusted, to gain or lose a certain per cent.

207. The price at which goods are bought is called the *first* or *prime cost;* that at which they are sold, the *selling price.* When the selling price is *greater* than the prime cost, there is a *gain;* when the selling price is *less* than the prime cost, there is a *loss.*

208. The gain or loss is always reckoned at a certain per cent. on the prime cost.

209. To find the gain or loss per cent., when the selling price and prime cost are known, —

RULE. *Divide the gain or loss by the prime cost, and the quotient, multiplied by* 100, *will be the gain or loss per cent.*

OBS. The gain or loss is the difference between the prime cost and the selling price.

The reason of this rule is obvious. The gain or loss, divided by the sum on which it was gained or lost, will give the gain or loss on 1 dollar, &c. This, multiplied by 100, will evidently be the gain or loss on 100, which will be the per cent.

1. What is the gain per cent. on cloth bought at $5 per yard and sold at $6 per yard ?

$$6 - 5 = 1. \quad \frac{1}{5} \times \frac{100}{1} = 20 \text{ per cent.}$$

The difference between the prime cost and selling

What is profit and loss? What is the prime cost? What is the selling price? On what is the gain or loss per cent. always reckoned? What is the rule for finding the gain or loss per cent., when the selling price and prime cost are known?

price is $1. This, divided by 5, and multiplied by 100, gives 20 per cent.

EXAMPLES.

2. Bought sugar at 6½ cents per pound, and sold it at 8 cents. What was the gain or loss per cent. ?

3. Bought cloth at $4.75 per yard, and sold it at $5 per yard. What was the gain or loss per cent. ?

4. Bought cloth at $5 per yard, and sold it at $4.75 per yard. What was the gain or loss per cent. ?

5. Bought molasses at 28 cents per gallon, and sold it at 31 cents per gallon. What was the gain or loss per cent. ?

6. Bought corn at 65 cents a bushel, and sold it at 62 cents per bushel. What was the gain or loss per cent. ?

7. Bought 7½ cords of wood at $3.50 per cord, and sold it at $3.75 per cord. What was the gain or loss per cent. ?

8. Bought butter at 17 cents per pound, and sold it at 20 cents per pound. What was the gain or loss per cent. ?

9. A broker bought stocks at $96 per share, and sold them at $102 per share. What did he gain per cent. ?

210. To find the prime cost, when the gain or loss per cent. and the selling price are known, —

RULE. *Divide the selling price by* 1 *plus the gain or minus the loss per cent., and the quotient will be the prime cost.*

10. A merchant sold cloth at $6 per yard, and gained 20 per cent. What was the prime cost ?

$$1+.20=1.20\ )\ 6.00\ (\ \$5$$
$$6.00$$

What is the rule for finding the prime cost, when the selling price and the gain per cent. are known ?

The selling price, $6, divided by 1+.20, which is 1.20, gives $5 as the prime cost.

EXAMPLES.

11. A farmer sold wood at $5.40 per cord, and gained 8 per cent. What was the prime cost?

12. A farmer sold hay at $14 per ton, and gained 25 per cent. What was the prime cost?

13. A merchant sold nails at 5½ cents per pound, and lost 10 per cent. What was the prime cost?

14. If 20 per cent. be gained on raisins, at $2.75 per box, what is the prime cost?

15. If 15 per cent. be gained on rice, at 4½ cents per pound, what is the prime cost?

16. If 12½ per cent. be gained on potatoes, at 48 cents per bushel, what is the prime cost?

17. If 9 per cent. be gained on cheese, at 10 cents per pound, what is the prime cost?

18. A merchant sold sugar at 6½ cents per pound, which was 10 per cent. less than it cost him. What was the prime cost?

19. A gentleman sold land at $175 per acre, which was 25 per cent. less than it cost him. What was the prime cost?

20. A merchant sold coal at $5½ per ton, which was 8 per cent. less than it cost him. What was the prime cost?

211. To find at what price goods must be sold, in order to gain or lose a certain per cent., —

RULE. *Find the required per cent. of the prime cost, and add this per cent. to it, if there is to be a gain; but subtract this per cent. from the prime cost, if there is to be a loss.*

What is the rule for finding the selling price to gain a certain per cent.?

21. Bought flour at $4 per barrel. At what price must it be sold to gain 20 per cent. ?

$$4 \times .20 = .80. \quad .80 + 4 = 4.80.$$

20 per cent. of $4 is 80 cents, which, added to $4, gives $4.80, the selling price.

EXAMPLES.

22. A merchant bought cloth at $5.00 per yard. At what price must it be sold, to gain 25 per cent. ?

23. A merchant bought sugar at 7½ cents per pound. At what price must it be sold, to gain 15 per cent. ?

24. A merchant bought molasses at 28 cents per gallon. At what price must it be sold, to gain 12½ per cent. ?

25. A merchant bought cotton at 9 cents per pound. At what price must it be sold, to lose 20 per cent. ?

26. A farmer bought apples at 42 cents per bushel. At what price must they be sold, to lose 25 per cent. ?

27. Bought cotton at $275 per bale. For how much must it be sold per bale, to gain 30 per cent. ?

28. A merchant bought flour at $4.50 per barrel. At what price must it be sold per barrel, to lose 15 per cent. ?

29. Bought molasses at 28 cents per gallon. At what price must it be sold per gallon, to lose 12½ per cent. ?

212. To find the gain or loss per cent. at any proposed price, when the selling price and the gain or loss per cent. is known, —

RULE. *First find the prime cost, and then the gain or loss per cent. at the proposed price.*

What is the rule for finding the gain or loss per cent. at any proposed price, when the selling price and the gain and loss are known?

EXAMPLES.

30. A merchant sold sugar at 8 cents per pound, and gained 10 per cent. What per cent. would he have gained if he had sold it at 9 cents per pound?

31. A farmer sold corn at 65 cents per bushel, and gained 5 per cent. What per cent. would he have gained if he had sold the corn at 70 cents per bushel?

32. A farmer sold rye at 95 cents per bushel, and gained 8 per cent. What would he have gained or lost per cent. if he had sold the rye at 80 cents per bushel?

33. A man sold his farm for $4560, which was 10 per cent. more than it cost him. What would he have gained or lost per cent. if he had sold it for $4000?

34. A farmer sold land at 5 cents per foot, and gained 25 per cent. more than it cost him. What would he have gained or lost per cent. if he had sold it at 3½ cents per foot?

35. A grocer sold tea at 45 cents per pound, and gained 10 per cent. What would he have gained per cent. if he had sold it at 50 cents per pound?

36. A merchant sold broadcloth at $4.75 per yard, and gained 12½ per cent. What would he have gained per cent. if he had sold it at $5.25 per yard?

37. A farmer sold oats at 37½ cents per bushel, and lost 14 per cent. What would he have gained or lost per cent. if he had sold them at 48 cents per bushel?

38. A merchant sold coffee at 11 cents per pound, and gained 10 per cent. What would he have gained per cent. if he had sold it at 12½ cents per pound?

39. A farmer sold potatoes at 35 cents per bushel, and lost 12½ per cent. What would he have gained or lost per cent. if he had sold them at 40 cents per bushel?

PRACTICAL QUESTIONS.

40. A man bought 12 acres of land at 3 cents per foot, and after keeping it 10 years, sells it at 20 per cent. advance. Allowing money to be worth 6 per cent., does he gain or lose, and how much?

41. A merchant bought 500 barrels of flour in Chicago, at $4 per barrel; he paid for freight to Boston 65 cents, and for truckage, 7 cents per barrel; he sold it in Boston at $5⅝ per barrel. How much did he gain per cent.?

42. A merchant bought at New Orleans 500 bales of cotton, of 300 pounds each, at 6¼ cents per pound; he paid for freight to Liverpool 1¼ cents per pound; for wharfage and truckage, $50; he sold the cotton for 9 cents per pound. Did he gain or lose, and how much per cent.?

43. A merchant bought in Maine 150,000 feet of lumber, at $9 per 1000 feet; he paid for freight, truckage, and wharfage, $4860. At what price per foot must he sell the lumber, to gain 20 per cent.?

44. A merchant bought in Vermont 6000 bales of wool, of 100 pounds each, at 40 cents per pound; he paid for freight and truckage to Boston 65 cents per bale. At what price per pound must he sell the wool, to gain 25 per cent.?

45. A merchant sold flour at $5½ per barrel, and thereby gained 12½ per cent. What would he have gained per cent. had he sold the flour at $6 per barrel?

46. A grocer bought 10 boxes of Havana sugar, of 400 pounds each, at 6¼ cents per pound; he paid for freight, truckage, &c., $1.75 per box; he gained 2 per cent. on the weight of the sugar; he sells it at 7½ cents per pound. How much does he gain per cent.?

47. A merchant sold tea at 45 cents per pound, and gained 12½ per cent. What would he have gained per cent. if he had sold the tea at 54 cents per pound?

RATIO.

SECTION XVIII.

213. RATIO is the relation which one quantity bears to another of the same kind with respect to magnitude.

214. *The ratio of two numbers is the quotient resulting from dividing the first by the second.* Thus the ratio of 12 to 4 is 3; the ratio of 30 to 5 is 6; since 12 divided by 4 is $\frac{12}{4}=3$, and 30 divided by 5 is $\frac{30}{5}=6$.

215. The two numbers are called the *terms* of the ratio. The first is called the *antecedent*, the second, the *consequent*, and may either be expressed in the form of a fraction, — the antecedent for the numerator, and the consequent for the denominator, — as $\frac{12}{4}$, or by placing two points between them, as 12 : 4.

OBS. Both of the numbers must either be abstract numbers or of the same kind.

216. *If both terms of the ratio be multiplied or divided by the same number, the ratio will not be changed.* Thus 12 : 4 is the same ratio as 6 : 2, 24 : 8, and 36 : 12. This is evident from the fact that the terms of a ratio are the terms of a fraction, which may be multiplied or divided without changing its value, (ART. 84.)

217. An *inverse* or *reciprocal* ratio is the ratio of the consequent to the antecedent, and is expressed by changing the order of the terms, or by inverting the fraction. Thus the ratio of 3 to 6, or $\frac{3}{6}$, is a direct ratio; the ratio of 6 : 3, or $\frac{6}{3}$, is an inverse or recipro-

---

What is ratio? What is the ratio of two numbers? What are the numbers called? How may a ratio be expressed? What is an inverse ratio?

cal ratio, and is always the same as the ratio of the reciprocals of those numbers. Thus $\frac{1}{6} : \frac{1}{3}$.

OBS. For definition of *reciprocal*, see Obs. 1, Art. 166.

218. A compound ratio is composed of two simple ratios. Thus,

The ratio of 5 : 20 is 4.
" " " 6 : 36 is 6.

The ratio of $5 \times 6 : 20 \times 36$ is 24.

As the terms of a ratio are the same as the terms of a fraction, they may be treated as such in every respect.

OBS. The question, What is the ratio of one number to another is the same as, What part of one number is another?

EXAMPLES.

1. What is the ratio of 8 : 4?
2. What is the ratio of 9 : 6?
3. What is the ratio of 12 : 16?
4. What is the ratio of 27 : 30?
5. What is the ratio of 39 : 48?
6. What is the ratio of 64 : 72?
7. What is the ratio of 84 : 104?
8. What is the ratio of 96 : 112?
9. What is the ratio of 102 : 28?
10. What is the ratio of 148 : 24?
11. What is the ratio of $\frac{1}{4} : 2$?
12. What is the ratio of $5 : \frac{3}{8}$?
13. What is the ratio of $\frac{3}{4} : \frac{4}{5}$?
14. What is the ratio of $\frac{7}{8} : \frac{9}{10}$?
15. What is the ratio of $5\frac{1}{2} : 16\frac{1}{2}$?
16. What is the ratio of $7\frac{1}{2} : 15\frac{1}{4}$?
17. What is the ratio of $3\frac{1}{4} : 19\frac{1}{5}$?
18. What is the ratio of $5\frac{2}{7} : 23\frac{3}{4}$?
19. What is the ratio of $7\frac{1}{8} : 11\frac{5}{7}$?

PROPORTION.

SECTION XIX.

219. PROPORTION is the union of two equal ratios. Thus, 6 : 12 : : 4 : 8, or $\frac{6}{12}=\frac{4}{8}$.

OBS. Proportion is expressed by four dots between the ratios. Thus, 2 : 4 : : 3 : 6.

220. The *first* and *fourth* terms are called the *extremes;* the *second* and *third*, the *means.*

OBS. The first and third terms are sometimes called the first and second antecedent; and the second and fourth, the first and second consequent.

221. In any proportion, if the first term be greater or less than the second, or equal to it, the third term will also be greater or less than the fourth, or equal to it.

222. *In every proportion the product of the extremes is equal to the product of the means.* Thus in the proportion, 20 : 5 : : 36 : 9, the product of $20\times9=180$, $5\times36=180$.

223. *Four numbers are in proportion when the product of the extremes is equal to the product of the means.*

224. If any three terms of a proportion are known, the other may easily be found.

225. *The product of the second and third terms, divided by the first, will give the fourth.*

OBS. Let the unknown term be represented by *u*.

$$8 : 6 : : 12 : u. \qquad \frac{\overset{3}{\not{6}}\times\overset{3}{\not{12}}}{\not{8}} = 9 = u.$$

What is proportion? What are the first and fourth terms called What are the second and third called?

$4 : 9 :: 7 : u. \quad \frac{9\times 7}{4} = 15\tfrac{3}{4} = u.$

226. *The product of the second and third terms, divided by the fourth, will give the first.*

$u : 4 :: 2 : 6. \quad \frac{2\times 4}{6} = 1\tfrac{2}{6} = u.$

$u : 8 :: 12 : 7. \quad \frac{12\times 8}{7} = 13\tfrac{5}{7} = u.$

227. *The product of the first and fourth terms, divided by the third, will give the second.*

$3 : u :: 12 : 18. \quad \frac{3\times 18}{12} = 4\tfrac{1}{2} = u.$

$6 : u :: 5 : 14. \quad \frac{14\times 6}{5} = 16\tfrac{4}{5} = u.$

228. *The product of the first and fourth terms, divided by the second, will give the third.*

$5 : 3 :: u : 15. \quad \frac{15\times 5}{3} = 25 = u.$

$7 : 10 :: u : 4. \quad \frac{7\times 4}{10} = 2\tfrac{8}{10} = u.$

229. If four quantities are in proportion, they will remain so if the extremes are put in the place of the means, and the means in the place of the extremes. The proportion will also remain the same if the means and the extremes are interchanged, as in the following examples:—

12 : 4 : : 15 : 5	4 : 12 : : 5 : 15
12 : 15 : : 4 : 5	4 : 5 : : 12 : 15
5 : 4 : : 15 : 12	15 : 12 : : 5 : 4
5 : 15 : : 4 : 12	15 : 5 : : 12 : 4

In each of these proportions the product of the extremes is equal to the product of the means.

As a proportion is composed of two simple ratios, it may always be changed to the form of fractions, and treated the same as simple ratios.

230. When three terms are given or known, they may be arranged in a proportion by the following rule : —

RULE. *Write that number for the third term which is of the same kind as the fourth term.*

If the fourth term must be greater than the third term, write the greater of the other two numbers for the second term, and the less for the first. If the fourth term must be less than the third, write the greater of the other two numbers for the first term, and the less for the second.

The product of the second and third terms, divided by the first, will give the fourth term.

OBS. 1. Factors which are common to the dividend and divisor should be cancelled.

OBS. 2. If either of the terms be a denominate number, they must be reduced to the lowest denomination mentioned in either.

OBS. 3. It may readily be perceived, from the conditions of the question, whether the fourth term be greater or less than the third.

1. If 12 yards of cloth cost $42, what will 16 yards cost?

$$12 : 16 : : 42 : u. \quad \frac{\overset{4}{\cancel{16}} \times \overset{14}{\cancel{42}}}{\cancel{12}} = 56.$$

Make $42, the same kind as the required or fourth term, the third term. It is evident that 16 yards must cost more than 12 yards; 16 yards must therefore be the second term, and 12 yards the first. The product of the second and third, divided by the first, gives $56. The 12 is cancelled, because its factors, 3 and 4, are also factors of 16 and 42.

What is the rule for finding the fourth term when the first three terms are known?

EXAMPLES.

2. If 6 yards of cloth cost $30, what will 9 yards cost?

3. If 12 bushels of wheat cost $15, how many bushels can be bought for $75?

4. If 14 pounds of flour cost 68 cents, what will 196 pounds cost?

5. If 9 cords of wood cost $30, how many cords can be bought for $156?

6. If 12 men can cut 49 cords of wood in a day, how many cords can 20 men cut in the same time?

7. If 12 pounds of rice cost $$\frac{2}{5}$, how many pounds can be bought for $$56\frac{1}{3}$?

8. If 24 acres cost $160, what will 164 acres cost?

9. If 12 men can perform a piece of work in 60 days, how many men would perform the same work in one third of the time?

10. If 48 men can build a wall in 36 days, how many men will be required to do the same in 72 days?

11. If 160 barrels of flour can be bought for $640, how many barrels can be purchased for $2240?

12. If 10 tons of hay can be purchased for $96, how many tons can be purchased for $240?

13. How far can a man travel in 16 days, if he travel 960 miles in 12 days?

14. If 9 men can reap 12 acres of rye in 15 days, how much would 15 men reap in the same time?

15. If A can do a piece of work in 9 days, and B can do the same in 12 days, what part of it can both do in 3 days?

16. If a person walk 396 miles in 14 days, of 12 hours each, in how many days, of 9 hours each, can he walk the same distance?

17. A hare, pursued by a dog, was 96 yards before him at starting. The dog ran 7 yards while the hare

ran 5. How far did the dog run before overtaking the hare?

18. $540 were divided between three persons, A, B, and C, in the following proportion: A received $5 as often as B received 6, and C 7. How much did each receive?

19. There are two numbers in proportion as 4 to 9, the larger of which is 117. What is the smaller?

20. There are two numbers in proportion as 9 to 7, the smaller of which is 126. What is the larger?

21. If a staff 3 feet long cast a shadow 5 feet in length, what is the height of a tower whose shadow at the same time is 175 feet?

22. If a reservoir containing 5690 gallons have two pipes, one of which discharges 50 gallons a minute, and the other admits 45 gallons a minute, in how long time, when it is full, will it be emptied?

23. If 36 yards of carpeting, $\frac{3}{4}$ of a yard wide, will cover a floor, how many yards, $1\frac{1}{4}$ wide, will be required to cover the same floor?

24. If $\frac{3}{4}$ of a ship be worth $12,000, how much will $\frac{7}{8}$ of the same be worth?

25. If $9\frac{1}{2}$ yards of cloth are worth $57, what are $12\frac{1}{4}$ yards worth?

26. If 16 men can perform a piece of work in 10 days, how many men can perform a piece of work 6 times as large in $\frac{1}{3}$ of the time?

27. If 9 men can mow 21 acres in 5 days, how many men would be required to mow 40 acres in the same time?

28. If A can cut 2 cords of wood in $12\frac{1}{3}$ hours, and B can cut 3 cords in $17\frac{1}{2}$ hours, how many cords can they both cut in $24\frac{1}{2}$ hours?

29. A, B, and C start at the same time and the same place to travel round an island 75 miles in circumference. A goes 5 miles a day, B 8, and C 10. In what time will they all be together?

30. If 69 yards of carpet, $\frac{3}{4}$ of a yard wide, will cover a floor 18 feet wide, what is the length of the room ?

31. At what time, between 3 and 4 o'clock, are the hands of a watch together ?

32. If a board be 9 inches wide, what must its length be to contain 6 square feet ?

33. Two ships sailed together from Boston to San Francisco. One sailed at the rate of 6 miles an hour, and the other $5\frac{1}{2}$, on an average, the whole distance. The first arrived in 165 days. In how many days did the other arrive ?

34. If the circumference of the wheel of a railroad car be 7 feet, and it make 5 revolutions in a second, in how long time will the car run from Boston to Providence, a distance of 42 miles ?

35. If the diameter of a wheel be 4 feet, and it make 445 revolutions in a mile, what would be the diameter of a wheel which makes $593\frac{1}{3}$ revolutions in the same distance ?

ANALYSIS.

231. The preceding examples may also be performed by analysis.

232. *Analysis* is a process of finding the value of a unit, or any part of a unit, of a given number, and from this, finding the value of any proposed number of units or parts of units. Thus, if 3 yards of cloth cost $12, what will 5 yards cost ? If 3 yards cost $12, 1 yard will cost $\frac{1}{3}$ of $12, or $4. If 1 yard cost $4, 5 yards will cost 5 times $4, which is $20.

36. If $2\frac{1}{2}$ bushels of corn cost $1.50, what will $7\frac{1}{2}$ bushels cost ? In $2\frac{1}{2}$ bushels there are 5 halves of a bushel. If 5 halves of a bushel cost $1.50, 1

What is analysis ?

half will cost 30 cents. If 1 half cost 30 cents, $7\frac{1}{2}$, which is 15 halves, will cost 15 times 30 cents, which is $4.50.

233. When the numbers are large, cancellation may be used with advantage in the analytical method.

37. If 15 cords of wood cost $52.50, how much will 20 cords cost?

$$\frac{\overset{17.50}{\cancel{52.50}}}{\cancel{15}} \times \frac{\overset{4}{\cancel{20}}}{1} = 70.00.$$

If 15 cords of wood cost $52.50, 1 cord will cost $\frac{1}{15}$ of $52.50, which may be expressed by writing $52.50 for a numerator, and 15 for a denominator. 20 cords will cost 20 times as much, which may be expressed by a fraction and by the sign of multiplication. The factors of 15 are 3 and 5, which are common to the numerator, 52.50, and to 20.

COMPOUND PROPORTION.

SECTION XX.

234. Compound Proportion is composed of a *compound* and a *simple* ratio.

235. A compound proportion may be resolved into as many simple proportions as there are pairs of terms, or may be combined into one.

236. In every compound proportion, one of the given numbers is always of the same kind as the required number.

237. Rule. *Write that number for the third term*

What is compound proportion? Into what may a compound proportion be resolved? What is of the same kind as the required term? What is the rule for compound proportion?

which is of the same kind as the required term. Then take two numbers of the same kind for the first and second terms, and arrange them as in simple proportion. Proceed in this manner with each pair of similar terms, and write them under the former. The continued product of the third and all the second terms, divided by the continued product of all the first terms, will give the required term.

1. If 16 horses eat 9 bushels of oats in 6 days, how many horses will eat 24 bushels in 8 days?

$$\begin{array}{l} 9 : 24 : : 16 \\ 8 : \ \ 6 : : \end{array} \qquad \frac{\overset{8}{\cancel{24}} \times \overset{2}{\cancel{16}} \times \overset{2}{\cancel{6}}}{\cancel{8} \times \cancel{9}} = 32.$$

Make 16 the third term, as it is of the same kind as the required term. As more horses are required to eat 24 bushels than 9, in the same time, make 24 the second term, and 9 the first, of the two remaining terms. Make 8 days the first, and 6 days the second term, as a less number of horses is required to eat the same quantity in 8 days than in 6 days. The continued product of all the second and the third terms, divided by the continued product of the first terms, gives 32 horses as the required term.

Obs. 1. The numbers to be multiplied and divided should be written as directed in the rule for cancellation, (Art. 65.)

Obs. 2. The first and second terms must be of the same denomination, and if the third term be a denominate number, it must be reduced to the lowest denomination mentioned in it.

Obs. 3. The conditions of every question should be thoroughly understood before attempting to arrange the numbers in proportion.

EXAMPLES.

2. If 6 cows produce 56 gallons of milk in 4 days, how many gallons will 26 cows produce in 7 days?

3. If 5 furnaces consume 30 tons of coal in 6 days, how many tons, at the same rate of consumption, will 6 furnaces consume in 40 days?

4. If a person travel 120 miles in 4 days, by walking 9 hours a day, what time will be required to travel 386 miles, by walking 7 hours a day?

5. If the freight of 2 tons, 12 cwt. 20 lb. for 42 miles be $8.50, what would be the freight of 16 tons, 10 cwt. 16 lb. for 140 miles?

Questions in compound proportion may be most intelligibly performed by analysis and cancellation.

6. If 32 horses eat 24 bushels of oats in 8 days, how many bushels will 16 horses eat in 6 days?

$$\frac{\overset{3}{\cancel{24}}\times\cancel{16}\times6}{\underset{2}{\cancel{32}}\times\cancel{8}}=\frac{18}{2}=9\text{ bushels.}$$

If 32 horses eat 24 bushels in 8 days, 1 horse will eat $\frac{1}{32}$ of 24 in 8 days; and in 1 day 1 horse will eat $\frac{1}{8}$ as much as in 8 days. 16 horses will eat 16 times as much as 1 horse in 1 day, and 6 times as much more in 6 days. 24 is therefore divided by 32 and 8, and multiplied by 16 and 6. The common factors being cancelled, the required term is 9 bushels.

7. If 6 men earn $150 in 4 weeks, working 6 days a week, how much would 10 men earn in 7 weeks, working 5 days a week?

8. If 84 men mow 72 acres of grass in 15 days, how many acres will 96 men mow in 12 days?

9. If 27 men build 54 rods of wall in 26 days, how many rods will 32 men build in 39 days?

10. If $300 gain $16 in 12 months, what principal will gain $10 in 9 months?

11. If 18 compositors can set up 24 sheets in 8 days, how many sheets would 45 compositors set up in 14 days?

12. If 5 persons can be maintained 30 days for $60,

how much money would be required to support 24 persons 365 days?

13. If 12 men can build 18 rods of wall in 30 days, working 9 hours a day, how many days would be required for 18 men, working 8 hours a day, to build a similar wall, 52 rods long?

14. If 16 men can build a wall 40 rods long, 4 feet high, and 3 feet thick, in 16 days, working 8 hours a day, in how many days will 20 men, working 9 hours a day, build a similar wall, 160 rods long, 6 feet high, and 5 feet thick?

15. If 1080 bricks, 8 inches long, and 2 inches wide, are required for a walk 20 feet long, and 6 feet wide, how many bricks will be required for a walk 100 feet long, and 4 feet wide?

PARTNERSHIP.

SECTION XXI.

238. PARTNERSHIP is the process of ascertaining the gain or loss of partners in trade.

239. The money invested is called the *capital* or *stock.*

240. The profit to be divided is called the *dividend.*

241. To find each partner's share of the profit or loss, when there is no reference to time, —

RULE. *Divide the whole gain or loss by the amount of the whole stock, or the number of shares in trade, and multiply the quotient by each man's stock, or share of the stock.*

What is partnership? What is the money invested called? What is called the dividend?

It is evident that the whole gain or loss, divided by the sum on which it was gained or lost, will give the gain or loss on $1. This, multiplied by each partner's stock or interest, will give the gain or loss of each partner.

1. Two persons, A and B, trade together. A puts in $600, B $800. They gain in one year $280. What is each man's share of the profit?

$$600 + 800 = 1400.$$

$$\frac{\overset{20}{\cancel{280}}}{\cancel{1400}} \times \frac{\overset{6}{\cancel{600}}}{1} = \$120, \text{ A's share.}$$

$$\frac{\overset{20}{\cancel{280}}}{\cancel{1400}} \times \frac{\overset{8}{\cancel{800}}}{1} = \$160, \text{ B's share.}$$

The whole gain, $280, divided by the amount of the whole stock, $1400, and multiplied by each partner's share, gives $120 for A's share, and $160 for B's share.

EXAMPLES.

2. Three persons, A, B, and C, enter into partnership. A advances $400, B $500, and C $600. They gain by trade $640. What is each person's share of the profit?

3. A, B, C, and D, purchase a ship. A pays for 6 shares, B for 5, C for 4, and D for 3. They receive net freight $4560. What sum ought each to receive?

4. An insolvent debtor owes to one of his creditors $450, to another $560, to a third $840. His property amounts to only $1250. How much will each of his creditors receive?

5. A gentleman left his estate in his will to his four sons, A, B, C, and D, as follows: To A $1600, to B $1500, to C 1860, and to D 2000. But his whole

property, after his debts were paid, amounted only to \$4840. How much will each son receive?

6. A, B, and C enter into partnership. A's stock was \$800, B's \$900, and C's \$1000. They lose 10 per cent. of the whole stock. What was each man's share of the loss?

7. Two persons engage in trade. The whole sum invested is \$5000. They gain \$600. A puts in $\frac{2}{3}$ of $\frac{5}{6}$ of $\frac{3}{4}$ of the whole, and B puts in the remainder. What was each man's share of the gain, and what did each put in?

8. A, B, and C owned a ship and cargo, valued at \$75,000, and insured for \$60,000, which was a total loss. A owned $\frac{1}{3}$, B $\frac{1}{4}$, and C the remainder. How much of the insurance ought each to receive, and what was each man's share of the loss?

9. A and B purchase a lot of land for \$4500. A pays $\frac{1}{3}$ of the price, B the remainder. They gain by the sale of it 20 per cent. of the first cost. What is each man's share of the gain?

10. A, B, and C purchase a lot of woodland. A pays \$200, B \$300, and C \$400. A works 40 days in cutting the wood, B 30 days, and C 20 days. A pays for cutting \$75, B \$50, and C \$20. They sell 400 cords of wood, at \$4 a cord. Allowing each man's labor to be worth \$1 a day, what ought each man to receive?

11. The sum of \$5000 is to be divided among 4 persons as follows: A is to receive $\frac{1}{5}$, B $\frac{1}{6}$, and C is to receive \$5 as often as D receives \$4. How much ought each man to receive?

242. To find each partner's share of the gain or loss, when the capital is invested for different periods, —

RULE. *Multiply each man's stock by the time of*

What is the rule for finding each partner's share of the gain or loss, when the capital is invested for different periods?

its continuance in trade, and divide the whole gain or loss by the sum of the several products, and multiply the quotient by the product of each man's stock and time, which will be each partner's gain or loss.

OBS. The principle of this rule may be applied to the solution of a great variety of questions, whose conditions are similar in their nature.

12. Three merchants, A, B, and C, entered into partnership. A put in \$600 for 4 months, B \$750 for 3 months, and C \$900 for 5 months. They gained \$488. What is each man's share of the gain?

$$\begin{aligned} 600 \times 4 &= 2400 \\ 750 \times 3 &= 2250 \\ 900 \times 5 &= \underline{4500} \\ & \quad 9150 \end{aligned}$$

$$\frac{488}{9150} = \frac{4}{\not{7}\not{5}} \times \frac{\overset{32}{\not{2}\not{4}\not{0}\not{0}}}{1} = \$128, \text{ A's share.}$$

$$\frac{4}{\not{7}\not{5}} \times \frac{\overset{30}{\not{2}\not{2}\not{5}\not{0}}}{1} = \$120, \text{ B's share.}$$

$$\frac{4}{\not{7}\not{5}} \times \frac{\overset{60}{\not{4}\not{5}\not{0}\not{0}}}{1} = \$240, \text{ C's share.}$$

Dividing the whole gain, 488, by the sum of the products, gives the fraction $\frac{488}{9150}$, which, reduced to its lowest terms, is $\frac{4}{75}$. This, multiplied by the product of each man's stock and time, gives \$128 for A's share, \$120 for B's share, and \$240 for C's share. 75 being a factor of 2400, 2250, and 4500, the numerators of the several fractions, and also in the denominator, it can be cancelled in each.

EXAMPLES.

13. A, B, and C entered into partnership. A put in $1000 for 4 months, B $900 for 5 months, C $1200 for 3 months. They lost $600. What was each partner's share of the loss?

14. A, B, and C contract to perform a certain piece of work. A employs 40 men for 4½ months, B 45 men for 3½ months, C 50 men for 2¼ months. They gain, after paying all expenses, $850. What part of the gain belongs to each?

15. A commenced business January 1, with a capital of $10,000. April 1 he admits B as a partner, with a capital of $5000. September 1 they receive C as a partner, with a capital of $3000. At the end of the year they had gained $2600. What was each man's share of the gain?

16. Four men hired a pasture for $175. A puts in 16 cows for 8 months, B puts in 12 cows for 9 months, C puts in 10 cows for 10 months, D puts in 8 cows for 12 months. How much ought each to pay?

17. Three persons, A, B, and C, form a partnership for 1 year, commencing January 1. A puts in $4000, B $3000, and C $2500. April 1, A withdraws $500 and B withdraws $600. June 1, C puts in $800 more. September 1, A furnishes $700 more, and B $400 more. At the end of the year they find they have gained $1500. What is each person's share of the gain?

18. A commenced business on the first day of January with a capital of $25,000. On the first of April he admits B as a partner, who furnishes $5000 capital. On the first of June they admit C as a partner, with a capital of $9000. At the end of two years they dissolve partnership, and find they have gained $8000. Each partner received his share of the stock and profits. What did each receive?

EQUATION OF PAYMENTS.

SECTION XXII.

243. Equation of Payments is the process of determining the average time for the payment of several sums, due at different periods.

244. Rule. *Multiply each sum by the time before it becomes due, and divide the sum of the products by the sum of the payments.*

Obs. 1. This rule has been considered by many as *incorrect*; but it is as *correct* as any rule founded upon *simple interest* can be.

If A owes B $200, $100 of which are to be paid in cash, and the other $100 to be paid in two years, the average time for the payment of the whole, according to the rule, would be *one year*, which is correct; for it is evident that if the money be paid according to the agreement, B would have at the end of two years $200, and the interest of $100 for two years. If the whole be paid in 1 year, B would have at the end of two years $200, and the interest of $200 for one year, which is the same as the interest of $100 for two years.

Obs. 2. If any sum be paid on the day from which the time is reckoned, it will have no product, but it must be added with the others in finding the sum of the payments.

Obs. 3. It is customary with merchants to add three days grace.

This rule is founded upon the principle that the use of $1 for any number of days is equivalent to the use of as many dollars as there are days for 1 day. Thus, the use of $1 for 150 days is equivalent to the use of $150 for 1 day.

1. A gentleman owes $150, payable in 60 days, $200, payable in 90 days, $400 in 120 days. At what time ought the whole to be paid at once?

$$\begin{array}{rl} 150\times\ \ 60= & 9000 \\ 200\times\ \ 90= & 18000 \\ 400\times 120= & 48000 \\ \hline 750 & 75000 \end{array} \qquad \frac{75000}{750}=100 \text{ days.}$$

What is equation of payments? What is the rule?

EXAMPLES

2. A owes B $500, of which $100 are to be paid in 4 months, $200 in 6 months, and the remainder in 9 months. What is the average time of payment?

3. A merchant sold the following bills of goods, on a credit of 6 months: May 10, a bill of $600; June 12, a bill of $450; September 20, a bill of $900. At what time will the whole become due?

4. A gentleman has to pay $5000 as follows: $540 in 4 months, $2400 in 8 months, $600 in 10 months, the remainder in 6 months. What is the average time for the payment of the whole sum?

5. A owes B $1200. $400 are due at the present time, $300 are due in 8 months, and $500 are due in 12 months. At what time ought the whole to be paid?

6. A gentleman left his son $1500, to be paid as follows: $\frac{1}{3}$ in 3 months, $\frac{1}{3}$ in 4 months, $\frac{1}{4}$ in 6 months, and the remainder in 8 months. At what time ought the whole to be paid at once?

7. A merchant bought goods amounting to $6000. He agrees to pay $500 in cash, $600 in 6 months, $1500 in 9 months, and the remainder in 10 months. At what time ought he to pay the whole in one payment?

8. A owes B $1600, to be paid in 6 months. A agrees to pay $600 in cash. At what time ought the remainder to be paid?

9. A gentleman purchased a farm for $3600, and agrees to pay $600 down, and the remainder in 5 equal semiannual instalments. At what time may the whole be paid at once?

10. A owes B $5000, $\frac{1}{4}$ to be paid in 30 days, $\frac{1}{3}$ in 3 months, $\frac{1}{4}$ in 70 days, and the remainder in 9 months. At what time ought the whole to be paid at once?

11. A bought goods of B, amounting to $1200. $\frac{1}{6}$ of the bill was to be paid in cash, $\frac{1}{4}$ to be paid in 2 months, the remainder in 9 months. At what time ought the whole to be paid?

245. To find the equated time of payments, when the sums due have different dates, —

RULE. *Find the time when each sum becomes due, and multiply each sum by the time between it and the first sum that is due, and divide the amount of the several products by the amount of the sums due.*

OBS. As the time at which the first sum is due is the period from which the average time is computed, the first sum will have no product.

The principle of this rule is the same as that of the preceding, (ART. 244.)

12. What is the equated time for the payment of the following sums: —

January 10,	due	$1000	
" 15,	"	2000 × 5 =	10000
" 25,	"	2500 × 15 =	37500
February 26,	"	1000 × 47 =	47000
March 27,	"	1500 × 76 =	114000
April 6,	"	2000 × 86 =	172000

10000) 380500 ($38\frac{1}{20}$
30000
80500
80000

38 days from January 10 is February 17, the equated time.

EXAMPLES.

13. A merchant sold 484 barrels of rosin, as follows: —

February 6,	4 months' cr.,	35 barrels,	@ $3.12½.
March 12,	"	38 "	@ $3.00.
" "	"	411 "	@ $2.62½.

What is the equated time for the payment of the whole?

What is the rule to find the equated time of payments, when the sums due have different dates?

14. A merchant sold 1650 barrels of flour, as follows: —

May 6,	3 months' cr.,	150 barrels,	@	$4.50.
" 20,	4 " "	400 "	@	4.75.
July 10,	5 " "	500 "	@	5.00.
August 4,	4 " "	600 "	@	4.25.

What is the equated time for the payment of the whole?

15. A merchant sold 576 barrels of rosin, as follows: —

May 3,	6 months' cr.,	62 barrels,	@	$2.50.
" 10,	" " "	100 "	@	2.50.
" 18,	cash,	10 "	@	2.50.
" 26,	30 days' cr.,	50 "	@	2.75.
" "	6 months' "	345 "	@	2.50.
" "	" " "	9 "	@	2.00.

What is the equated time for the payment of the whole?

TAXES.

SECTION XXIII.

246. A TAX is a certain sum of money assessed on individuals to pay the expenses of government, or of a corporation, society, district, &c.

247. Taxes for paying the expenses of government are usually assessed on the property of individuals, and on male persons without property, when of a certain age. The individual tax is called the *poll tax*.

OBS. 1. *Poll* is from a Dutch word, signifying the *head*.

OBS. 2. Every male individual over 20 years of age, is required by the laws of Massachusetts to pay a poll tax.

OBS. 3. Taxable property is of two kinds. viz., *real estate* and *personal property*. All property that is unchanging, such as houses, lands, &c., is called *real estate*. All other property, such as money,

What is a tax? How are taxes usually assessed?

notes, stocks, mortgages, furniture, cattle, &c., is called *personal property.*

248. To assess a state or other tax, —

Rule. *Find the amount of all the taxable property, real and personal, and also the number of individuals liable to be taxed. If there be a poll tax, multiply the number of individuals or polls by the amount assessed on each poll, and subtract the product from the sum to be raised. Divide the remainder by the whole amount of taxable property, which will be the tax on one dollar. Multiply the amount of each man's property by the tax on one dollar, and the product will be the tax on his property, which, added to his poll tax, will be his whole tax.*

1. A tax of $12,500 is to be assessed upon the inhabitants of a town in which there are 840 ratable polls. The tax on each poll is $1.50. The real and personal property in the town is valued at $5,364,560. What will a man's tax be, whose real estate is valued at $8500, and personal property at $15,000?

$12,500—$1260 for poll taxes=$11,240. $\frac{11240}{5364560}$ =.002095 on a dollar. This sum, multiplied by the amount of his property, produces $49.2325+$1.50, the poll tax,=$50.7325, the required sum.

EXAMPLES.

2. What is a man's tax whose real estate is valued at $25,000, personal $12,500, and 3 polls, in a town where the property is valued at $4,278,560, and the amount assessed is $10,000, there being 750 ratable polls?

3. What will be the amount of a man's tax in the same town, whose property is valued at $75,000?

4. What will be the amount of a man's tax in the same town, whose property is valued at $7320.47?

What is the rule to assess a state or other tax?

DUTIES.

SECTION XXIV.

249. The sum of money imposed by government on most goods imported from foreign countries is called *duties*.

250. Duties are either *ad valorem* or *specific*.

251. *Ad valorem* duties are a certain per cent. of the value or first cost of the goods.

Obs. *Ad valorem* is from the Latin, signifying *according to the value*.

252. *Specific* duties are a certain sum imposed on the gallon, yard, hundred weight, ton, &c.

Obs. All duties are regulated by government, and are often changed. By the tariff of 1846, ad valorem duties are required to be paid.

253. It is usual to make certain deductions, called *tare*, *draft*, &c., before specific duties are imposed.

254. *Tare* is a deduction of a certain per cent. of the weight of the goods, made after the draft, for the box, cask, &c.

255. *Draft* is a deduction of a certain number of pounds made for waste.

256. An allowance of 2 per cent. is made for leakage on any liquor in casks, subject to duty by the gallon, and 10 per cent. on all ale, beer, and porter, in bottles, and 5 per cent. on other liquor in bottles; or the duty is computed on the actual quantity, at the option of the importer, to be ascertained by actual measurement.

257. All goods entered at the custom-house are, when bought and sold, subject to the same deductions as are made when they are entered.

What are duties? What are ad valorem duties? What are specific? What is tare? What is draft? What allowances are usually made?

258. On all goods not entered at the custom-house there is a special agreement respecting tare, draft, &c.

259. To find the ad valorem duty on any merchandise, —

RULE. *Multiply the amount of the first cost of the merchandise by the rate per cent.*

OBS. In estimating the first cost of goods, all expenses in the foreign port are included.

EXAMPLES.

1. What is the ad valorem duty, at 30 per cent., on glass, china, and stone ware, the invoice amounting to $1260 ?
2. What is the ad valorem duty, at 30 per cent., on an invoice of cutlery, amounting to $760 ?
3. What is the ad valorem duty, at 100 per cent., on an invoice of brandy and cordials, amounting to $1560 ?
4. What is the ad valorem duty, at 30 per cent., on a cargo of Russia iron, amounting to $7560 ?
5. What is the ad valorem duty, at 40 per cent., on an invoice of raisins, figs, dates, and currants, amounting to $4560 ?

OBS. An invoice is a written account of merchandise, containing the price of each article and the charges of exportation.

260. To find the specific duty on any merchandise, —

RULE. *First make all the required deductions for tare, draft, &c., and multiply the remainder by the duty on each gallon, yard, pound, &c.*

6. What is the specific duty, at 2 cents per pound, on 950 bags of coffee, each weighing 200 pounds, tare 2 per cent.?
7. What is the specific duty, at 12 cents per gallon, on 25 pipes of molasses, each pipe containing 120 gallons, allowing 2 per cent. for leakage ?

What is the rule for ad valorem duties ? for specific ?

ALLIGATION MEDIAL.

SECTION XXV.

261. Alligation Medial is the process of finding the price of a mixture composed of several ingredients of different values.

262. To find the price of a mixture, when the price of each ingredient and the quantity are given, —

Rule. *Multiply each ingredient by its price, and divide the sum of the products by the sum of the ingredients. The quotient will be the price of the mixture.*

1. A grocer mixed 10 pounds of tea worth 60 cents per pound with 15 pounds worth 75 cents per pound, and 25 pounds worth 40 cents per pound. What is the price of one pound of the mixture?

$$\begin{aligned} 10\times60 &= \ 6.00 \\ 15\times75 &= 11.25 \\ 25\times40 &= 10.00 \\ \hline 50 \qquad & \quad 27.25 \end{aligned}$$

$$27.25 \div 50 = 54\tfrac{1}{2} \text{ cents.}$$

EXAMPLES.

2. A goldsmith mixes 2 lbs. of gold of 18 carats fine, 3 lbs. of 20 carats fine, and 4 lbs. of 22 carats fine. How many carats fine is the mixture?

3. If 16 bushels of oats at 40 cents per bushel, 10 bushels of corn at 65 cents per bushel, and 12 bushels of barley at 75 cents per bushel, were mixed together, what would be the price of the mixture?

4. A grocer mixes 10 pounds of tea at 40 cents per pound, 20 pounds at 45 cents per pound, 30 pounds at 50 cents per pound. What would a pound of this mixture be worth?

What is alligation medial? What is the rule?

ALLIGATION ALTERNATE.

SECTION XXVI.

263. ALLIGATION ALTERNATE is the process of finding what quantity of ingredients, of different prices, must be taken to compose a mixture of a given price.

RULE. *Find how much is gained or lost by taking one of each kind of the proposed ingredients. Then take one or more of the ingredients, or such parts of them, as will make the gain and loss equal.*

1. A trader would mix four sorts of tea, viz., at 4 s., 6 s., 9 s., and 12 s. per pound. How many pounds of each sort must be taken, that the mixture may be worth 8 s. per pound?

		Gain.	Loss.
1 lb. at 4,	8 — 4 =	4	
1 lb. at 6,	8 — 6 =	2	
1 lb. at 9,	9 — 8 =		1
1 lb. at 12,	12 — 8 =		4
1 lb. at 9,	9 — 8 =		1
		6	6

By taking 1 pound of each ingredient, it is evident there will be a gain of 1 s. By taking 2 pounds at 9 s., the gain and loss will be equal. The same result may be obtained by taking $1\frac{1}{4}$ pounds at 12 s. and 1 pound of each of the other ingredients.

EXAMPLES.

2. How much sugar, at 5 cents, at 6 cents, and 9 cents per pound, must be mixed together, that the mixture may be worth 7 cents per pound?

3. In what proportion must sugar, at 4 cents, at 8

What is alligation alternate? What is the rule?

cents, and 10 cents per pound, be mixed, that the com pound may be worth 6 cents per pound?

4. In what proportion must gold, valued at 12, 16, 18, and 24 carats fine, be mixed, that the compound may be worth 20 carats fine?

5. In what proportion must grain, valued at 50 cents, 56 cents, 62 cents, and 75 cents per bushel, be mixed together, that the compound may be 62 cents per bushel?

6. In what proportion must corn at 70 cents per bushel, oats at 45 cents per bushel, and rye at 60 cents per bushel, be mixed, that the compound may be worth 55 cents per bushel?

264. When one or more of the ingredients are limited in quantity, to find the other ingredients, —

RULE. *Find how much is gained or lost by taking one of each of the proposed ingredients, in connection with the ingredient which is limited, and if the gain and loss be not equal, take such of the proposed ingredients, or such parts of them, as will make the gain and loss equal.*

7. How much sugar that is worth 6 cents, 10 cents, and 13 cents per pound, must be mixed with 20 pounds of sugar worth 15 cents per pound, to make a compound worth 11 cents per pound?

		Gain.	Loss.	
1 lb. at 6,	11 — 6 =	5		
1 lb. at 10,	11 — 10 =	1		
1 lb. at 13,	13 — 11 =		2	
20 lb. at 15,	15 — 11 =		80	
		6	82	— 6 = 76
15 lb. at 6,	11 — 6 =	75		
1 lb. at 10,	11 — 10 =	1		
		82	82	

16 lb. at 6 cents, 2 lb. at 10, and 1 lb. at 13 cents.

What is the rule when one or more of the ingredients are limited?

By taking one of each of the ingredients at their prices, in connection with the limited ingredient, 20 lbs., there is a loss of 76 cents. As there is 5 cents gain on 1 lb. at 6 cents, on 15 lbs. there would be a gain of 75 cents. And as there is 1 cent gain on 1 lb. at 10 cents, by taking 15 lbs. more at 6, and 1 lb. at 10, making 16 lbs. of the one, and 2 lbs. of the other, the gain and loss are equal.

EXAMPLES.

8. How much sugar, at 10 cents, at 9 cents, and 8 cents per pound, must be mixed with 30 pounds at 6 cents per pound, that the compound may be worth 8 cents per pound?

9. How much gold of 16 and 18 carats fine must be mixed with 90 ounces of 22 carats fine, that the compound may be worth 20 carats fine?

265. To find the quantity of each ingredient, when the sum of the ingredients and the mean price are given, —

RULE. *Find the least quantity of each ingredient by* ART. 263. *Then divide the given amount by this sum, and multiply the quotient by the quantities found for the least proportional quantities.*

10. How much tea, at 25 cents, 35 cents, 50 cents, and 70 cents per pound, must be mixed with 18 pounds, that the mixture may be worth 45 cents?

		Gain.	Loss.
1 lb. at 25,	45 — 25 =	20	
1 lb. at 35,	45 — 35 =	10	
1 lb. at 50,	50 — 45 =		5
1 lb. at 70,	70 — 45 =		25
4		30	30

180 ÷ 4 = 45 lb. of each.

What is the rule when the sum of the ingredients and the mean price are given?

Proof.

$45 \times 25 = 1125$
$45 \times 35 = 1575$
$45 \times 50 = 2250$
$45 \times 70 = 3150$

$8100 \div 180 = 45$

EXAMPLES.

11. How much sugar, at 6 cents, 8 cents, 10 cents, and 12 cents per pound, must be mixed with 200 pounds, that shall be worth 9 cents per pound?

12. How much gold, of 18, 20, and 22 carats fine, must be mixed with alloy, in order to form a composition of 40 ounces, worth 16 carats fine?

DUODECIMALS.

SECTION XXVII.

266. DUODECIMALS are a species of denominate numbers. 12 of each lower denomination make one of the next higher.

267. The denominations are *feet*, *inches* or *primes*, *seconds*, *thirds*, *fourths*, &c., which are distinguished from each other by marks called *indices*.

Table.

12′	inches	=	1	foot.
12″	seconds	=	1′	inch or prime.
12‴	thirds	=	1″	second.
12⁗	fourths	=	1‴	third.

1 ft.	=	12′	inches.
1 ft.	=	144″	seconds.
1 ft.	=	1728‴	thirds.
1 ft.	=	20736⁗	fourths.

What are duodecimals? What are the denominations, and how are they distinguished? Recite the table.

OBS. *Duodecimals* is from a Latin word, signifying *twelve*.

268. Duodecimals may be added or subtracted in the same manner as other denominate numbers.

EXAMPLES.

1. What is the sum of 14 ft. 10 in. 8″ 9‴, 15 ft. 6 in. 7″ 10‴, 17 ft. 11 in. 4″ 6‴?
2. What is the sum of 20 ft. 10 in. 7″ 9‴, 26 ft. 6 in. 5″ 4‴, 30 ft. 5 in. 2″ 3‴, 40 ft. 8 in. 9″ 3‴?
3. From 24 ft. 7 in. 3″ 4‴ take 18 ft. 8 in. 6″ 5‴.
4. From 60 ft. 6 in. 8″ 9‴ take 56 ft. 7 in. 9″ 11‴.

MULTIPLICATION OF DUODECIMALS.

269. RULE. *Write the corresponding denominations of the multiplicand and multiplier under each other. Multiply each denomination in the multiplicand in succession by each denomination in the multiplier, and for every twelve in the product add one to the next product, and write the remainder under its corresponding denomination. Add together the partial products, and for every twelve add one to the next column.*

270. The foot being considered the unit or whole number, the multiplication of duodecimals is the same in principle as the multiplication of common fractions. Thus 6 ft. $\times$ 6 ft. $=$ 36 square feet. 6 ft. $\times$ 6 in. is the same as 6 ft. $\times \frac{6}{12}$ ft. $= \frac{36}{12} =$ 36 in. 6 ft. $\times$ 6″ is the same as 6 ft. $\times \frac{6}{144} = \frac{36}{144} =$ 36″. 6 ft. $\times$ 6‴ is the same as 6 ft. $\times \frac{6}{1728} = \frac{36}{1728}$ ft. $=$ 36‴. And 6 in. $\times$ 6 in. is the same as $\frac{6}{12}$ ft. $\times \frac{6}{12} = \frac{36}{144}$ ft. $=$ 36″. 6 in. $\times$ 6″ is the same as $\frac{6}{12} \times \frac{6}{144} = \frac{36}{1728}$ ft. $=$ 36‴. 6″ $\times$ 6″ is the same as $\frac{6}{144}$ ft. $\times \frac{6}{144} = \frac{36}{20736}$ ft. $=$ 36 ⁗, &c.

How may duodecimals be added and subtracted? What is the rule for the multiplication of duodecimals?

It is evident from the preceding illustration that—
Feet × by feet produce sq. ft.
Feet × by inches produce inches.
Feet × by seconds produce seconds.
Feet × by thirds produce thirds, &c.
Inches × by inches produce seconds.
Inches × by seconds produce thirds.
Inches × by thirds produce fourths.
Seconds × seconds produce fourths, &c.

Obs. 1. The product of any two denominations will have as many indices as there are in both of the denominations.

Obs. 2. Many mechanics divide the foot into tenths, hundredths, &c.; and it is to be regretted that this is not the universal practice, as the calculations would be much more simple and easy.

5. Multiply 10 ft. 6 in. 8″ by 8 ft. 4 in. 7″.

ft.	in.	″	‴	⁗
10	6	8		
8	4	7		
84	5	4		
3	6	2	8	
	6	1	10	8
ft. 88	5′	8″	6‴	8⁗

ft.	in.	″	‴	⁗
10	6	8		
8	4	7		
	6	1	10	8
3	6	2	8	
84	5	4		
ft. 88	5′	8″	6‴	8⁗

Obs. It is obvious from the preceding examples that it is not important whether the lowest denomination in the multiplicand be multiplied first by the highest or lowest denominations in the multiplier, provided the remainder be placed directly under its corresponding denomination.

EXAMPLES.

6. Multiply 15 ft. 9 in. 8″ by 14 ft. 6 in. 9″.

7. How many square feet in a board 12 ft. 4 in. long and 10 inches wide?

8. How many square feet in a board 15 ft. 4 in. long and 1 ft. 3 in. wide?

9. How many cubic feet of wood in a load 8 ft. 3 in. long, 3 ft. 9 in. high, and 3 ft. 10 in. wide?

10. How many cubic feet in a pile of wood that is

50 ft. 9 in. long, 4 ft. 3 in. high, and 3 ft. 10 in. wide?

11. How many square feet in a room that is 16 ft. 4 in. long and 18 ft. 8 in. wide?

12. How many cubic feet in a block of granite 15 ft. 6 in. long, 3 ft. 4 in. wide, and 2 ft. 9 in. thick?

13. How many cubic feet in a load of gravel that is 5 ft. 4 in. long, 3 ft. 9 in. wide, and 11 in. high?

14. How many cubic feet in a cellar that is 40 ft. 9 in. long, 36 ft. 8 in. wide, and 6 ft. 4 in. deep?

15. How many cubic feet in a stone wall that is 3 ft. 4 in. high, 2 ft. 9 in. thick, and 120 ft. long?

16. How many yards of carpeting, a yard wide, will be required to cover a floor that is 24 ft. 6 in. long and 18 ft. 6 in. wide?

17. How many yards of carpeting, $\frac{3}{4}$ yard wide, will be required to cover a floor that is 15 ft. 4 in. long and 16 ft. 6 in. wide?

18. How many feet of wood in a pile that is 60 ft. 9 in. long, 4 ft. 3 in. high, and 3 ft. 10 in. wide?

Obs. Cellars and other excavations are generally estimated by squares, 216 cubic feet being considered a square.

19. How many squares in a cellar that is 50 ft. 6 in. long, 36 ft. wide, and 6 ft. 4 in. deep?

20. How many squares in a trench that is 4 ft. deep, 3 ft. 6 in. wide, and 100 ft. long?

21. How many square yards of plastering in a room 16 ft. 8 in. long, 15 ft. 6 in. wide, and 9 ft. 10 in. high?

22. How many bushels of corn will a bin hold that is 20 ft. long, 4 ft. 6 in. wide, and 3 ft. 8 in. high, there being 2150.4 cubic inches in a bushel?

23. How many hogsheads will a cistern contain that is 13 ft. 6 in. long, 9 ft. 8 in. wide, and 8 ft. high, allowing 231 cubic inches to a gallon, and 63 gallons to a hogshead?

24. In a room 16 ft. 8 in. long, 15 ft. wide, and 10 ft. high, there are three doors, each of which is 6 ft. 8 in. high and 2 ft. 8 in. wide; two windows 3 ft. 4 in. wide and 5 ft. 6 in. high; and a fireplace that is 4 ft. high and 4 ft. 6 in. wide. How many yards of plastering are there in the room, and what would it cost to plaster it, at $12\frac{1}{2}$ cents per yard?

25. How many squares of gravel would be required to raise a lot of land 5 ft. 6 in. that is 100 ft. long and 80 ft. wide; and how many horse-loads, supposing each load to be 5 ft. 6 in. long, 4 ft. 3 in. wide, and 10 in. high?

INVOLUTION.

SECTION XXVIII.

271. Involution is the process of finding the required power of any number, by multiplying it into itself.

272. The number multiplied is called the *root.*

273. *Powers* are of different orders, as the *second*, the *third*, the *fourth*, &c. The second power is also called the *square*, the third the *cube*, the fourth the *biquadrate.* In the second power the root is used twice as a factor, in the third power it is used three times, in the fourth four times, &c.

274. The power of a number and the number of times it is taken as a factor is generally indicated by a small figure called an *index*, or *exponent*, placed above the given number, a little to the right. The index of the second power is 2, and of the third power is 3, &c.

$$5^2 = 5 \times 5 = 25$$
$$5^3 = 5 \times 5 \times 5 = 125$$
$$5^4 = 5 \times 5 \times 5 \times 5 = 625$$

What is involution? What is the root of a number?

275. To find the required power of any number, —

RULE. *Multiply the given number into itself in succession, till it is taken as a factor as many times as there are units in the index of the required power.*

OBS. A common fraction is raised to any required power by raising each term to the required power. A mixed number should first be reduced to an improper fraction, or to a decimal, and compound and complex fractions to simple ones.

EXAMPLES.

1. What is the second power of 6?
2. What is the second power of 9?
3. What is the third power of 4?
4. What is the third power of 7?
5. What is the third power of 8?
6. What is the fourth power of 3?
7. What is the fourth power of 6?
8. What is the fifth power of 12?
9. What is the sixth power of 9?

Table of Roots and Powers.

Roots.	1	2	3	4	5	6	7
2d Power.	1	4	9	16	25	36	49
3d Power.	1	8	27	64	125	216	343
4th Power.	1	16	81	256	625	1296	2401
5th Power.	1	32	243	1024	3125	7776	16807
6th Power.	1	64	729	4096	15625	46656	117649
7th Power.	1	128	2187	16384	78125	279936	823543

276. *The product of two powers of the same number is equal to that number raised to the power denoted by the sum of the indices.* Thus, $6^3 \times 6^4 = 6^{3+4} = 6^7$. $6^3 = 216$; $6^4 = 1296$. $1296 \times 216 = 279936 = 6^7$.

277. *The quotient of any power of a number, divided by a less power of the same number, is equal to*

What is the rule for finding the required power of any number?

the number raised to the power denoted by the difference of their indices. Thus $6^7 \div 6^4 = 6^{7-4} = 6^3$.

$$6^7 = 279936. \quad 6^4 = 1296.$$
$$279936 \div 1296 = 216 = 6^3.$$

278. *Any power of a composite number is equal to the product of the same powers of its factors.* Thus, $12^3 = 1728$; $4^3 \times 3^3 = 1728$.

EVOLUTION.

SECTION XXIX.

279. EVOLUTION is the reverse of involution, and is the process of finding the roots of any given numbers.

280. The square root of any number is designated by the following signs: $\sqrt{\ }$, $(\ )^{\frac{1}{2}}$. Thus, $\sqrt{49}$ and $(49)^{\frac{1}{2}}$ denote that the square root of 49 is to be extracted. $\sqrt{49}$ or $(49)^{\frac{1}{2}} = 7$. $\sqrt[3]{\ }$ or $(\ )^{\frac{1}{3}}$ denote that the cube root is to be extracted.

281. The root of any number is a factor which, being multipled into itself a given number of times, will produce that number. The name of a root indicates the number of times a factor is taken. Thus, in the square root a factor is taken twice, in the cube root three times, &c.

OBS. 1. When the root of a number can be exactly found, it is called a *perfect power*, and the root a *rational number;* but if the root cannot be extracted without a remainder, it is called an *imperfect power*, and its root an *irrational number*, or *surd*.

OBS. 2. A number may be a perfect power of one degree, but an imperfect power of another degree. Thus 36 is a perfect power of the second degree, but an imperfect power of any other degree.

OBS. 3. Every power and root of 1 is 1.

What is evolution? How is the square root of any number designated? How is the cube, &c.? What is the root of any number?

SQUARE ROOT.

SECTION XXX.

282. THE SQUARE ROOT of any number is that number which, being multiplied into itself, produces the given number.

283. RULE. *Separate the given number into periods of two figures each, beginning with units.*

Find the greatest square in the left hand period, and write its root in the quotient. Subtract the square from the left hand period, and to the remainder annex the next period for a partial dividend.

Double the figure of the root for a trial divisor. Find how many times this divisor is contained in the partial dividend, omitting the right hand figure, and write the result in the quotient, and also annex it to the trial divisor for a complete divisor. Multiply the complete divisor by the last figure, and subtract the product from the partial dividend, and to the remainder annex the next period for a new partial dividend.

To the complete divisor add the last figure of the root for a new trial divisor. Divide as before, till all the periods are annexed.

OBS. 1. The square root of a common fraction may be found by extracting the root of the numerator and denominator, when it can be done without a remainder; when it cannot, reduce the fraction to a decimal, and then extract the root.

OBS. 2. When the trial divisor is greater than the partial dividend, a cipher must be written in the root, and also annexed to the trial divisor.

OBS. 3. If there be a remainder, periods of two ciphers each may be annexed.

What is the square root? What is the rule?

The principle on which this rule depends may be illustrated as follows: —

The square of 45 is 2025.

```
 45          45 is equal to 4 tens and 5 units.
 45
----
 25  the square of the units.
 20  the product of the tens and units.
 20  the product of the tens and units.
16   the square of the tens.
----
2025
                              40+ 5
                              40+ 5
                             -------
                             200+25
                        1600+200
                        ------------------
                        1600+400+25=2025
```

$40^2=1600 \quad 2\,(40\times5)=400 \quad 5^2=25$

$1600+400+25=2025$

284. From the preceding illustration it is evident that *the square of the sum of two numbers is equal to the square of the two numbers, plus twice their product, or to the square of the tens, plus the square of the units, plus twice the product of the tens by the units.*

285. It is also evident that *any square may be separated into two other squares, plus twice the product of their roots.* Thus 2025 may be separated into 1600, the largest square it contains, and 425. 425 will contain another square, plus twice the product of the roots of the two squares. Dividing 425 by twice 40, the root of the first square, gives 5 for the root of the other square; since $2\times40\times5+5^2=425$.

Obs. 1. If a number consists of three or more figures, the last figure may be considered as the units, and the other figures as the tens.

What is the square of two numbers?

OBS. 2. The trial divisor is always twice the figures of the root previously found, or twice the tens. The complete divisor is always the same as the trial divisor, plus the units.

OBS. 3. The reason for omitting the right hand figure in the partial dividend is, that the trial divisor is always considered as tens.

1. What is the square root of 205209?

```
        205209 ( 453
        16
        ----
  85    452
   5    425
  ---   -----
  90     2709
 903     2709
        -----
```

Separate the given numbers into periods of two figures each. Find the greatest square in the left hand period, 20, which is 16. The root of 16 is 4, which write in the quotient. Annex the next period, 52. Double the figure in the root for a trial divisor, which is 8. Find how many times 8 is contained in the partial dividend, omitting the right hand figure, 45. 8 is contained in 45, 5 times. Write 5 in the quotient, and also annex it to the trial divisor, 8, making 85 for the complete divisor. Multiply the complete divisor, 85, by 5, and subtract from 452. To the remainder, 27, annex the next period, 09, making 2709, for a new partial dividend. Add the last figure, 5, to 85, for a trial divisor. Find how many times the trial divisor, 90, is contained in the partial dividend, omitting the right hand figure. Write the result, 3, in the quotient, and annex it to the trial divisor, for a new complete divisor. Multiply the complete divisor by 3, and subtract from the partial dividend. As there is no remainder, and as there are no more periods, the solution is finished.

OBS. The given numbers are divided into periods of two figures each, because no square number can have more figures than twice the number of figures in the root, nor but *one* less. The largest square of any one figure is 81. $9 \times 9 = 81$. The square of 1 is 1.

EXAMPLES.

2. What is the square root of 9216?
3. What is the square root of 27225?
4. What is the square root of 119025?
5. What is the square root of 717409?
6. What is the square root of 59049?
7. What is the square root of 205209?
8. What is the square root of 94249?
9. What is the square root of 18671041?
10. What is the square root of 34421689?
11. What is the square root of 38875225?
12. What is the square root of 1048576?
13. What is the square root of 924524836?
14. What is the square root of 640.894864?
15. What is the square root of 744.326?
16. What is the square root of 983?
17. What is the square root of 1728?
18. What is the square root of 6?
19. What is the square root of 15?
20. What is the square root of 24?
21. What is the square root of $\frac{25}{36}$?
22. What is the square root of $\frac{49}{63}$?
23. What is the square root of $\frac{64}{81}$?
24. What is the square root of $30\frac{1}{4}$?
25. What is the square root of $272\frac{1}{4}$?
26. What is the square root of $\frac{1296}{2304}$?
27. What is the square root of $\frac{1020}{3026}$?
28. What is the square root of $\frac{3}{7}$ of $\frac{21}{25}$?
29. What is the square root of $51\frac{21}{25}$?
30. What is the square root of $56\frac{1}{4}$?
31. What is the square root of $61\frac{36}{49}$?
32. What is the square root of 42852312064?
33. What is the square root of 203461696?
34. What is the square root of .00032754?
35. What is the square root of .00001296?
36. What is the square root of .00002025?
37. What is the square root of .00003025?

286. *The difference of the squares of two numbers, divided by their sum, will give their difference.*

287. *The difference of the squares of two numbers, divided by their difference, will give their sum.*

EXAMPLES.

38. The difference of the squares of two numbers is 24; their sum is 12. What are the numbers? (See ART. 42.)

39. The difference of the squares of two numbers is 128; their sum is 16. What are the numbers?

40. The difference of the squares of two numbers is 7191; their sum is 141. What are the numbers?

41. The difference of the squares of two numbers is 80; their difference is 4. What are the numbers?

42. The difference of the squares of two numbers is 3471; their difference is 39. What are the numbers?

288. *Twice the product of two numbers, added to the sum of their square, will be the square of their sum.* Thus, $6\times5\times2=60.60+36+25=121=$ square of $6+5$.

EXAMPLES.

43. Complete the square of 16+ +9.
44. Complete the square of 144+ +25.
45. Complete the square of 625+ +16.
46. Complete the square of 729+ +81.
47. Complete the square of 3136+ +625.

289. *The square of a number and twice the product of this number by a second being known, the other may be found by dividing twice their product by twice*

What will the difference of the squares of two numbers, divided by their sum, give? Divided by their difference? Twice the product of two numbers, added to the sum of their square, will be equal to what? When the square of one number and twice the product of this number by a second are known, how may the second number be found?

the root of the first number. Thus, 36+24. Dividing 24 by 2×6, the root of 36, gives 2 for the second number. By adding the square of 2 to 36+24+4, completes the square.

EXAMPLES.

48. Complete the square of 81+72+.
49. Complete the square of +12+36.
50. Complete the square of 144+168+.
51. Complete the square of 729+270+.
52. Complete the square of 1296+864+.
53. Complete the square of +2250+2025.

290. *When the means of a proportional are equal, either is called the mean proportional between the extremes.* Thus, 4 : 6 : : 6 : 9 is a proportion, (Art. 219,) and the 6 is a mean proportional between 4 and 9.

291. To find a mean proportional between two given numbers,—

Rule. *Multiply one number by the other, and extract the square root of their product.*

54. What is the mean proportional between 4 and 9?

$\sqrt{(4\times9)}=6$, the mean proportional.

EXAMPLES.

55. What is the mean proportional between 4 and 16?

56. What is the mean proportional between 3 and 27?

57. What is the mean proportional between 9 and 49?

58. What is the mean proportional between 16 and 144?

59. What is the mean proportional between 25 and 64?

How is a mean proportional found?

60. What is the mean proportional between 19 and 41?

61. What is the mean proportional between 12 and 20?

62. What is the mean proportional between 36 and 48?

63. What is the mean proportional between 5 and 60?

64. If an article weighed in one scale of a false balance be 4 lb., and, when put in the other, 9 lb., what was its true weight?

65. If a quantity of sugar weighed in one scale of a false balance be 9 lb., and, in the other, 16 lb., what was the true weight of the sugar?

PRACTICAL QUESTIONS.

66. The area of a circle is 7085 inches. What is the side of a square whose area is equal to the circle?

67. A general wishes to form an army of 50176 men into a square. What would be the number in rank and file?

68. A gentleman has a field measuring 6 acres, 2 rood, 24 rods. How many rods long would a square field be, that should contain 4 times as much?

69. What must be the length of a square field containing 20 acres?

292. A triangle is a figure having three sides and three angles. When one side is perpendicular to another side, it is a *right-angled triangle*, and the angle between the two sides is a right angle. The longest side, or the one opposite the right angle, is called the *hypothenuse*, and the other two sides the *base* and *perpendicular*.

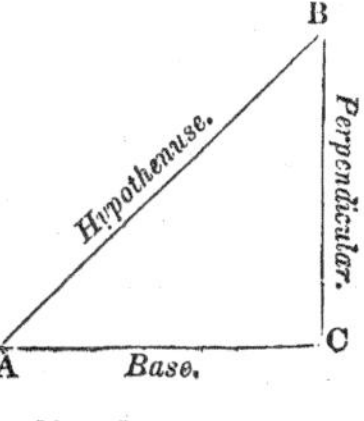

293. *The square described on the hypothenuse, or*

longest side, is equal to the sum of the squares described on the other two sides. Thus, suppose the longest side is 10 ft., the base 6 ft., and the perpendicular 8 ft. $10^2=100$. $6^2=36$. $8^2+36=100$.

294. Any two sides of a right-angled triangle being known, the other side may easily be found.

295. To find the hypothenuse, when the base and perpendicular are known, —

RULE. *Add the square of the base and the square of the perpendicular, and the square root of their sum will be the hypothenuse.*

70. If the base of a right-angled triangle be 12 ft., and the perpendicular 16 ft., what is the hypothenuse?

$$12^2=144. \quad \sqrt{(16^2+144)}=\sqrt{400}=20.$$

296. To find the base, when the perpendicular and hypothenuse are known, —

RULE. *Subtract the square of the perpendicular from the square of the hypothenuse, and the square root of the remainder will be the base.*

71. If the hypothenuse of a right-angled triangle be 30 ft., and the perpendicular 24 ft., what is the base?

$$30^2=900. \quad 24^2=576. \quad \sqrt{(900-576)}=\sqrt{324}=18.$$

297. To find the perpendicular, when the base and hypothenuse are known, —

RULE. *Subtract the square of the base from the square of the hypothenuse, and the square root of the remainder will be the perpendicular.*

72. If the hypothenuse be 15 ft., and the base 9 ft., what is the perpendicular?

$$15^2=225. \quad 9^2=81. \quad \sqrt{(225-81)}=\sqrt{144}=12.$$

OBS. The preceding principles may be represented by the following formula: $AB^2 = AC^2 + BC^2$, $AB^2 - AC^2 = BC^2$, $AB^2 - BC^2 = AC^2$.

73. If the hypothenuse of a right-angled triangle be 80 ft., and the base 48 feet, what is the perpendicular?

74. If the hypothenuse be 90 ft., and the perpendicular 72 ft., what is the base?

75. If the base be 30 ft., and the perpendicular 40 ft., what is the hypothenuse?

76. A ladder 40 ft. long will reach a window 32 ft. high, on one side of a street, and on the other side another window, 24 ft. high. What is the breadth of the street?

77. The wall of a castle, 45 ft. high, is surrounded with a ditch; and a ladder 75 ft. long will reach from the outside of the ditch to the top of the wall. What is the breadth of the ditch?

78. Two ships sail from the same port; one of them sails due south 96 miles; the other 72 miles due east. How far are they then distant?

79. The wall of a castle is surrounded with a ditch 60 ft. broad, and requires a ladder 75 ft. long to reach from the outside of the ditch to the top of the wall. What is the height of the wall?

80. A gentleman owns a farm in the form of a square, containing 250 acres. What is the length of one side of the farm, and what is the distance between the opposite corners?

81. There was a liberty pole, whose top, being broken partly off, touched the ground at 15 feet distance from the foot of the pole. The broken piece was 25 feet. What was the whole length of the pole before being broken?

82. Suppose the top of a liberty pole, 180 feet high, to be broken off in part, so that the top reached the ground at the distance of 60 feet from the pole. What is the length of the piece broken off? (ARTS. 42, 286.)

CUBE ROOT.

SECTION XXXI.

298. The Cube Root of any number is that number which, multiplied into itself twice, will produce the given number.

299. Rule. *Separate the given number into periods of three figures each, beginning with units.*

Find the greatest cube in the left hand period, and write the root in the quotient. Subtract the cube from the left hand period, and annex the next period for a partial dividend.

Multiply the square of the root by 3, *and annex two ciphers for the first trial divisor.*

Find how many times the trial divisor is contained in the partial dividend, and write the result in the quotient.

Complete the divisor by adding to the trial divisor three times the tens of the root, with a cipher annexed, multiplied by the last figure, plus the square of the last figure.

Multiply the complete divisor by the last figure of the root, and subtract the product from the partial dividend, and annex the next period.

Add to the last complete divisor the difference between the preceding trial and complete divisor, plus the square of the last figure, and annex two ciphers for the succeeding trial divisor.

Obs. 1. If there be a remainder, periods of three ciphers each may be annexed. If there are decimals, they must be separated into periods of three figures each, beginning with tenths.

Obs. 2. If the trial divisor be not contained in any partial dividend, place a cipher in the quotient, and annex two ciphers to the trial divisor for the next trial divisor, and annex the next period.

Obs. 3. The given numbers are separated into periods of three figures each, because the largest cube of any one figure is expressed by three figures.

What is the cube root? What is the rule?

300. The root of every cube may be considered as divided into tens and units. If there be but two figures in the root, the first figure will be the tens, and the last figure the units. If there be more than two figures in the root, the last figure will be the units, and the preceding figures the tens. Thus, 45 may be separated into 4 tens and 5 units; 4532 may be separated into 453 tens and 2 units.

301. *The trial divisor is always three times the square of the tens, or the figures of the root, previously found, omitting the last figure;* as, $3(40)^2=4800$.

302. *The complete divisor is always three times the square of the tens, plus three times the tens, multiplied by the units, plus the square of the units;* as, $3(40)^2+3(40\times5)+5^2=5425$.

303. *The difference between any trial divisor and the next complete divisor is three times the tens, multiplied by the units, plus the square of the units;* as, $3(40\times5)+5^2=625=5425-4800$.

304. *Three times the square of any number is three times the square of the tens of that number, plus six times the tens, multiplied by the units, plus three times the square of the units;* as, $3(40+5)^2=3(40)^2+6(40\times5)+3(5)^2=6075$.

It is evident, from the preceding propositions, that three times the tens multiplied by the units, plus the square of the units, added to any trial divisor, will give the next complete divisor.

It is also evident, that the difference between the complete divisor and the preceding trial divisor, plus the square of the units, added to the complete divisor, will give the next trial divisor.

What is the trial divisor? What is the complete divisor? What is the difference between the trial and complete divisors? What is three times the square of any number?

Obs. The principle of the rule may be expressed and demonstrated by the following formula. Let a = the tens, b = the units.

$$(a+b)^3 = a^3 + 3\,a^2\,b + 3\,a\,b^2 + b^3 = \text{any cube.}$$

$$\begin{array}{rl}
3\,a^2 = & \text{the first trial divisor.} \\
3\,a^2 + 3\,a\,b + b^2 = & \text{the complete divisor.} \\
3\,a\,b + b^2 = & \text{difference between the trial and complete divisor.} \\
b^2 & \\
\hline
3\,a^2 + 6\,a\,b + 3\,b^2 = & \text{succeeding trial divisor.}
\end{array}$$

$$3\,(a+b)^2 = a^2 + 2\,a\,b + b^2 \times 3 = 3\,a^2 + 6\,a\,b^2 + 3\,b^2$$

When there is more than one figure in the root, they may be expressed by $a + b$. The trial divisor will then be three times the square of $a + b$, which is $3\,a + 6\,a\,b + 3\,b^2$. The same result may be obtained by adding to the preceding complete divisor $3\,a^2 + 3\,a\,b + b^2$, the difference between it and the preceding trial divisor, $3\,a\,b + b^2$ plus b^2.

$$\left.\begin{array}{r} 3\,a^2 + 3\,a\,b + b^2 \\ 3\,a\,b + b^2 \\ + b^2 \end{array}\right\} = 3\,a^2 + 6\,a\,b + 3\,b^2$$

This rule may be illustrated as follows: 45 are 4 tens and 5 units, which may be raised to the cube thus:—

$$\begin{array}{r}
40+5 \\
40+5 \\
\hline
200+25 \\
1600+200 \\
\hline
1600+400+25 \\
40+5 \\
\hline
8000+2000+125 \\
64000+16000+1000 \\
\hline
64000+24000+3000+125=91125
\end{array}$$

$64000=(40)^3$. $24000=3\,(40^2\times 5)$. $3000=3\,(40\times 5^2)$. $125=(5^3)$.

It is obvious that the cube of any number, of more than one figure, is *the cube of the tens, plus three times the square of the tens, multiplied by the units, plus three times the tens, multiplied by the square of the units, plus the cube of the units.*

1. What is the cube root of 91125?

$$
\begin{array}{rr}
 & 91125\ (\ 40+5=45 \\
(40)^3= & 64000 \\
3\,(40)^2=4800 & \\
3\,(40\times 5)+5^2=\ \ 625 & 27125 \\
5425\times 5= & 27125
\end{array}
$$

2. What is the cube root of 93082856768?

$$
\begin{array}{rrcl}
 & & & 93.082,856,768\ (\ 4532 \\
 & (4)^3 & = & 64 \\
4^2\times 3= & 4800,\text{ tr. div.} & & 29082,\text{ par. div.} \\
40\times 3=120\times 5+5^2= & 625,\text{ dif.} & & \\
\text{com. div.} & 5425\times 5 & = & 27125 \\
5^2= & 25 & & 1957856,\text{ par. div.} \\
 & 607500,\text{ tr. div.} & & \\
450\times 3=1350\times 3+3^2= & 4059,\text{ dif.} & & \\
\text{com. div.} & 611559\times 3 & = & 1834677 \\
3^2= & 9 & & 123179768,\text{ par. div.} \\
 & 61562700,\text{ tr. div.} & & \\
4530\times 3=13590\times 2+2^2= & 27184,\text{ dif.} & & \\
\text{com. div.} & 61589884\times 2 & = & 123179768
\end{array}
$$

EXAMPLES.

3. What is the cube root of 2985984?
4. What is the cube root of 75686967?
5. What is the cube root of 644972544?
6. What is the cube root of 50243409?
7. What is the cube root of 12862247607?
8. What is the cube root of 163039787847?
9. What is the cube root of 50023150823736?
10. What is the cube root of 64240300125?
11. What is the cube root of 94996712418949125?
12. What is the cube root of 94997087172244118016?
13. What is the cube root of 163.04?
14. What is the cube root of 3.46?
15. What is the cube root of 3.375?

305. To extract the cube root of a common fraction, —

Rule. *Find the cube root of the numerator and the denominator, if it can be done without a remainder; if not, reduce the fraction to a decimal, and then find the nearest cube root.*

EXAMPLES.

16. What is the cube root of $\frac{27}{64}$?
17. What is the cube root of $\frac{512}{729}$?
18. What is the cube root of $\frac{72}{243}$?
19. What is the cube root of $91\frac{1}{8}$?
20. What is the cube root of $\frac{4}{5}$?
21. What is the cube root of $\frac{1}{2}$?

306. To find two mean proportionals between two given numbers, —

Rule. *Divide the greater number by the less, and extract the cube root of the quotient. The less number multiplied by this root will be the least mean proportional. The larger number divided by this root will be the greatest mean proportional.*

22. What are the two mean proportionals between 6 and 162?

$162 \div 6 = 27$. $\sqrt[3]{27} = 3$. $3 \times 6 = 18$, the least mean proportional; $162 \div 3 = 54$, the greatest.

EXAMPLES.

23. What are the two mean proportionals between 11 and 704?
24. What are the two mean proportionals between 12 and 2592?
25. What are the two mean proportionals between 8 and 1000?

What is the rule for finding two mean proportionals between two given numbers?

26. What are the two mean proportionals between 9 and 3087?

27. What are the two mean proportionals between 25 and 12800?

ROOT OF HIGHER POWERS.

SECTION XXXII.

307. THE root of any power may be found by extracting in succession the roots denoted by the several factors of the index of any higher power. Thus, the fourth root of any number may be found by extracting the square root twice. $3^4=81$. The square root of 81 is 9. The square root of 9 is 3. The sixth root may be found by extracting the square root, and then the cube root; as, $6=2\times3$; the eighth root, by extracting the square root three times; as, $8=2\times2\times2$. The ninth root, by extracting the cube root twice; as, $9=3\times3$. The twelfth, by extracting the square root twice, and then the cube; as, $12=2\times2\times3$.

308. To extract the root of any power, —

RULE. *Separate the given number into periods of as many figures each as there are units in the index of the power, beginning with units.*

Find the first figure of the root, and subtract its power from the left hand period, and to the remainder annex the first figure of the next period, for a dividend.

Involve the root to the power next inferior to that denoted by the index, and multiply it by the index, for a divisor.

Find how many times the divisor is contained in

What is the rule for finding the root of higher powers?

the partial dividend, and the result will be the second figure of the root.

Involve the figures of the root already found to the power denoted by the index, and subtract it from the two left hand periods, and to the remainder annex the first figure of the next period, for a dividend, and find the divisor as before. Proceed in this manner till the whole root is found.

1. What is the fifth root of 13542593318343?

13542593318343 (423
4^5 = 1024, 1st div.

$4^4 \times 5$=1280 3302

42^5 = 130691232

42^4=3111696, 2d. div. 47347011, 2d dividend.
423^5 = 13542593318343

EXAMPLES.

2. What is the fourth root of 32015587041?

3. What is the seventh root of 2423162679857794-647?

4. What is the eighth root of 10249978135798471-35681?

ARITHMETICAL PROGRESSION.

SECTION XXXIII.

309. Arithmetical Progression is a series of numbers which increase or decrease by a *common difference;* as, 2, 4, 6, 8, 10, is an increasing series, in which the common difference is 2; 15, 12, 9, 6, 3, is a decreasing series, in which the common difference is 3.

310. The numbers which form the series are called

What is arithmetical progression?

the *terms* of the progression. The first and last terms are called the *extremes*, the other terms are called the *means*.

311. *When any three of the five following terms are known, the remaining two may be found.*

312. To find the common difference when the two extremes and the number of terms are known,—

RULE. *Divide the difference of the extremes by the number of terms, less one, and the quotient will be the common difference.*

This rule may be represented by the formula, thus: Let a=first term, d=common difference, l=last or required term, n=the number of terms, s=sum of all the terms.

$$d = \frac{l - a}{n - 1}$$

It is evident that the number of differences will be one less than the number of terms, as there will be one term at each end of the series. It is also evident that the whole difference between the extremes will be the amount of the differences between each term.

1. The extremes are 2 and 56, and the number of terms 19. Required the common difference.

$$\frac{56 - 2}{19 - 1} = \frac{54}{18} = 3 = \text{common difference.}$$

EXAMPLES.

2. If the extremes be 3 and 95, and the number of terms 24, what is the common difference?

3. If the extremes be 2 and 100, and the number of terms 30, what is the common difference?

4. A man starts on a journey, and travels the first

What is the rule for finding the common difference, when the extremes and number of terms are known?

day 5 miles, and increases his journey every day the same number of miles; on the twelfth day he travels 49 miles. What was the daily increase?

313. To find the sum of all the terms, when the two extremes and the number of terms are known, —

RULE. *Multiply half the sum of the extremes by the number of terms, and the product will be the sum of all the terms.*

This rule may be represented by the following formula: —

$$s = \frac{n\,a + n\,l}{2} \qquad s = \frac{n\,(a + l)}{2}$$

The reason of this rule will be obvious from the following illustration: If the terms of any series are written under each other, reversing the order, it will be seen that the sum of all the terms is equal to the sum of the extremes multiplied by half of the number of terms. Thus,

2	4	6	8	10	12	14	16	18
18	16	14	12	10	8	6	4	2
20	20	20	20	20	20	20	20	20

The extremes are 2 and 18. The sum of the extremes is 20. Half the sum multiplied by 9, the number of terms, gives 90, which evidently is the sum of the terms, as 180 is twice the sum of the terms.

EXAMPLES.

5. The first term of an arithmetical series is 5, the last term 56, and the number of terms 18. What is the sum of the series?

6. The first term is 8, the last term is 50, the number of terms 15. What is the sum of the series?

What is the rule for finding the sum of the terms, when the extremes and number of terms are known?

7. How many times does a clock strike in 24 hours?

8. What is the sum of the series of numbers, 1, 2, 3, 4, 5, &c., to 75?

9. If 100 stones are placed on the ground in a right line, at the distance of 1 yard from each other, and a basket 1 yard from the first stone, how far will a person have to walk, who, starting from the basket, shall bring them, one by one, to the basket?

314. To find the number of terms, when the extremes and common difference are known, —

RULE. *Divide the difference of the extremes by the common difference, and the quotient, increased by* 1, *will be the number of terms.*

This rule may be represented by the formulas, thus: —

$$n = \frac{l - a}{d} + 1 \qquad n = \frac{2\,s}{a + l}$$

EXAMPLES.

10. If the extremes are 4 and 145, and the common difference 3, what is the number of the terms?

11. A person going a journey travelled 6 miles the first day, and increased his journey every day by 3 miles; his last day's journey was 45 miles. How many days did he travel?

315. To find any required term, when the first term and common difference are known, —

RULE. *Multiply the number of terms, less one, by the common difference, and add this product to the first term, for the required term.*

What is the rule for finding the number of terms, when the extremes and common difference are known? What for finding any required term, when the first term and common difference are known?

OBS. If the series be descending, the product of the common difference and the number of the terms, less 1, must be subtracted from the first term.

This rule may be represented by the formulas : —

$$l = a + n\,d - d. \quad l = a + (n - 1)\,d. \quad l = \frac{2\,s}{n} - a.$$

12. The first term of a series is 3, the common difference 4, the number of terms 25. What is the last term ?

(25—1)×4+3=99, the term required.

EXAMPLES.

13. If the first term of a descending arithmetical series be 50, the common difference $2\frac{1}{2}$, and the number of terms 20, what is the last term ?

14. If a debt can be discharged in 30 weeks by paying $2 the first week, $5 the second, $8 the third, &c., what will be the last payment ?

15. What is the hundredth term of a series whose first term is 5, and the common difference $2\frac{1}{2}$?

GEOMETRICAL PROGRESSION.

SECTION XXXIV.

316. GEOMETRICAL PROGRESSION is a series of numbers which *increase* or *decrease* by a *common ratio*. Thus, 2, 4, 8, 16, &c., is an increasing series, in which the ratio is 2; and 81, 27, 9, 3, 1, $\frac{1}{3}$, $\frac{1}{9}$, &c., is a decreasing series, in which the ratio is 3.

317. The numbers forming the series are called the

What is geometrical progression? What is an increasing series? What is a decreasing series? What are the terms ?

terms of the progression. The first and last terms are called the *extremes*, the other terms the *means*.

318. In every geometrical series there are five things to be considered — the first term, the last term, the number of terms, the ratio, and the sum of all the terms. When any three of them are known, the other two may easily be found.

OBS. 1. The product of the extremes of any series is equal to the product of any two terms equally distant from them.

OBS. 2. The square of any term is equal to the product of any two terms equally distant from it.

319. To find the last term, or any required term, when the first term, the ratio, and the number of terms are known, —

RULE. *Multiply the first term by the ratio, raised to a power whose index is one less than the number of terms.*

1. The first term of a geometrical series is 2, the ratio 3, the number of terms 15. What is the last term?

$3^{14}=4782969$. $3^2=9$. $3^4=9\times9=81$. $3^8=81\times81=6561$. $3^{12}=6561\times81=531441$. $3^{14}=531441\times9=4782969$. $4782969\times2=9565938$, the last term.

EXAMPLES.

2. The first term of a geometrical series is 4, the ratio is 2, the number of terms 10. What is the last term?

3. The first term is 50, the ratio $\frac{1}{2}$. What is the tenth term?

4. The first term is 3, the ratio $\frac{1}{4}$. What is the sixth term?

What are the extremes? What are the means? What is the ratio? How many things are considered in a geometrical progression? What are they? How many must be known, to find the others? What is the rule for finding the last term?

5. The first term is 8, the ratio 3. What is the fifth term?

6. A bought a piece of cloth containing 16 yards. He agreed to pay 4 cents for the first yard, 8 cents for the second, and doubling the price for each succeeding yard. What did he give for the last yard?

7. If a farmer plant a grain of wheat which yields the first year 20 grains, and if each grain be planted, and yield in the same proportion for ten successive years, how many grains would he have at the end of the tenth year?

320. To find the sum of the series, when the first term, the last term, and the ratio are known, —

RULE. *Multiply the greater term by the ratio, and from the product subtract the less extreme. Then divide the remainder by the ratio less one, and the quotient will be the sum of the series.*

8. The first term of a geometrical series is 2, the ratio 3, the last term 9565938. What is the sum of the series?

$$(9565938\times3)-2\div(3-1)=14348906.$$

OBS. 1. If the series be decreasing, the last term multiplied by the ratio must be subtracted from the first term, and the remainder divided by 1 minus the ratio.

OBS. 2. The sum of an infinite series may be found by dividing the first term by 1, minus the ratio; the ratio in an infinite series always being considered a fraction.

$$s=\frac{a}{1-r}.$$

EXAMPLES.

9. The greatest term of a geometrical series is 26244, the number of terms 9, and the ratio 3. What is the sum of the series?

What is the rule for finding the sum of the series?

10. What is the fifth term of the series, 1, $\frac{1}{2}$, $\frac{1}{4}$, &c.?

11. A gentleman bought a horse on the condition that he should pay 1 cent for the first nail in his shoes, 2 cents for the second, 4 cents for the third, and in geometrical progression for each successive nail. The horse had four shoes, and 8 nails in each shoe. What did the gentleman pay for the horse?

12. What is the sum of the infinite series, $\frac{1}{2}$, $\frac{1}{4}$, $\frac{1}{8}$, $\frac{1}{16}$, $\frac{1}{32}$?

13. A lady, about to be married, agreed to receive as her portion $5 on her wedding day, and to have that sum doubled on each succeeding anniversary of her marriage for ten years. What was the lady's portion?

321. Compound interest may be computed by geometrical progression, the principal being the first term, the ratio the amount of $1 for 1 year, and the number of years, plus 1, the number of terms.

OBS. When payments are to be made *half yearly*, *quarterly*, &c., the number of terms will be *one more* than the number of payments.

14. What is the amount of $100 for 4 years, at 6 per cent. compound interest?

$100=$first term; $1.06=$ratio; $4+1=5=$number of terms; $1.06^4\times100=126.2476=$last term, or amount.

OBS. The amount of $1 may also be found by the table, (ART. 184.)

EXAMPLES.

15. What will $50 amount to in 6 years, at 5 per cent. compound interest?

16. What will be the amount of $1000 in 10 years, at 6 per cent. compound interest?

17. In what time will $500 be doubled, at 6 per cent. compound interest?

18. What will be the amount of $300 in 8 years, at $5\frac{1}{2}$ per cent. compound interest?

19. What will be the amount of $400 in 4 years, at 6 per cent. compound interest, to be paid quarterly?

ANNUITIES.

SECTION XXXV.

322. An Annuity is a periodical income, payable at equal intervals, such as yearly, half-yearly, quarterly, monthly, &c.

323. When annuities are to continue for a definite number of years, they are called *annuities certain;* but when they are to be paid only so long as one or more individuals shall live, they are called *contingent* or *life annuities.* When an annuity is to continue forever, it is called a *perpetuity.*

324. When an annuity has commenced, or commences immediately, it is said to be *in possession;* but when it does not commence till some time has elapsed, it is called an *annuity in reversion.*

325. When annuities are not paid when due, they are said to be *in arrears.*

326. To find the amount of an annuity at simple interest, —

Rule. *Find first the last term by* Art. 315, *and then the sum of the series by* Art. 313.

1. What will an annuity of $300 amount to in 5 years, at 6 per cent. simple interest?

$300+\overline{18\times4}=372$, the last term. $\dfrac{372+300}{2}\times 5=1680$, the sum.

EXAMPLES.

2. What will an annuity of $400 amount to in 8 years, at 5 per cent. simple interest?

3. What will an annuity of $600 amount to in 10 years, at 6 per cent. simple interest?

What is an annuity? What are annuities certain? What are contingent annuities? What is a perpetual annuity? What is an annuity in possession? What, in reversion? What, in arrears? What is the rule for finding the amount of an annuity at simple interest?

TABLE I.

Showing the amount of an annuity of 1 *dollar from* 1 *year to* 30.

Years.	4 per cent.	5 per cent.	$5\frac{1}{2}$ per cent.	6 per cent.
1	1.000000	1.000000	1.000000	1.000000
2	2.040000	2.050000	2.055000	2.060000
3	3.121600	3.152500	3.168025	3.183600
4	4.246464	4.310125	4.342266	4.374616
5	5.416322	5.525631	5.581091	5.637093
6	6.632975	6.801913	6.888051	6.975319
7	7.898294	8.142008	8.266894	8.393838
8	9.214226	9.549109	9.721573	9.897468
9	10.582795	11.026564	11.256259	11.491316
10	12.006107	12.577893	12.875354	13.180795
11	13.486351	14.206787	14.583498	14.971643
12	15.025805	15.917127	16.385590	16.869941
13	16.626838	17.712983	18.286798	18.882138
14	18.291911	19.598632	20.292572	21.015066
15	20.023588	21.578564	22.408663	23.275971
16	21.824531	23.657492	24.641139	25.672528
17	23.697512	25.840366	26.996402	28.212880
18	25.645413	28.132385	29.481205	30.905653
19	27.671229	30.539004	32.102671	33.759992
20	29.778078	33.065954	34.868318	36.785592
21	31.969202	35.719252	37.786075	39.992727
22	34.247970	38.505214	40.864309	43.392290
23	36.617888	41.430475	44.111846	46.995828
24	39.082604	44.501999	47.537998	50.815577
25	41.645908	47.727099	51.152588	54.864512
26	44.311745	51.113454	54.965979	59.156383
27	47.084214	54.669126	58.989109	63.705766
28	49.967583	58.402583	63.233510	68.528112
29	52.966286	62.322712	67.711353	73.639798
30	56.084938	66.438847	72.435478	79.058186

TABLE II.

Showing the present worth of an annuity of 1 dollar from 1 year to 30.

Years.	4 per cent.	5 per cent.	5½ per cent.	6 per cent.
1	0.96154	0.95238	0.94786	0.94339
2	1.88609	1.85941	1.84631	1.83339
3	2.77509	2.72325	2.69793	2.67301
4	3.62990	3.54595	3.50514	3.46511
5	4.45182	4.32948	4.27028	4.21236
6	5.24214	5.07569	4.99553	4.91732
7	6.00205	5.78637	5.68297	5.58238
8	6.73274	6.46321	6.33457	6.20979
9	7.43533	7.10782	6.95220	6.80169
10	8.11090	7.72173	7.53762	7.36009
11	8.76048	8.30641	8.09254	7.88687
12	9.38507	8.86325	8.61852	8.38384
13	9.68565	9.39357	9.11708	8.85268
14	10.56312	9.89864	9.58965	9.29498
15	11.11839	10.37966	10.03759	9.71225
16	11.65230	10.83777	10.46216	10.10589
17	12.16567	11.27407	10.86461	10.47726
18	12.65930	11.68959	11.24607	10.82760
19	13.13394	12.08532	11.60765	11.15812
20	13.59033	12.46221	11.95034	11.46992
21	14.02916	12.82115	12.27524	11.76408
22	14.45112	13.16300	12.58317	12.04158
23	14.85684	13.48857	12.87504	12.30338
24	15.24696	13.79864	13.15170	12.55036
25	15.62208	14.09394	13.41391	12.78336
26	15.98277	14.37518	13.66250	13.00317
27	16.32959	14.64303	13.89810	13.21053
28	16.66306	14.89813	14.12142	13.40616
29	16.98371	15.14107	14.33310	13.59072
30	17.29203	15.37245	14.53375	13.76483

27. To find the amount of an annuity at compound interest, in arrears, —

RULE. *Subtract* 1 *from the amount of* $1 *at compound interest for the given time, and divide the difference by the interest of* $1 *for a year, and multiply the quotient by the given annuity.*

OBS. When the payments are *half-yearly, quarterly, monthly*, &c., the amount of $1 for the time between the payments will be the ratio, and the number of payments the number of terms.

4. What is the amount of an annuity of $200, which has not been paid for 4 years, at 6 per cent. compound interest?

$1.06^4 - 1. = .262476.$ $(.262476 \div .06) \times 200 = 874.9200.$

OBS. The amount of an annuity may be found by multiplying the amount of $1, as found in the table, by the given annuity.

EXAMPLES.

5. What is the amount of a salary of $500 for 10 years, at 6 per cent. compound interest?

6. What is the amount of a salary of $1500 for 12 years, at 6 per cent. compound interest?

7. If the annual rent of a farm, which is $300, be not paid for 5 years, what will the rent amount to, at 6 per cent. compound interest?

328. To find the present worth of an annuity, —

RULE. *Subtract the present worth of* $1 *for the given time from* 1, *and divide the difference by the interest of* $1 *for one year, and multiply the quotient by the given annuity; or multiply the present worth of an annuity of* $1 *at the given rate and time by the given annuity.*

OBS. The present worth of an annuity may be found by multiplying the given annuity by the present worth, as found in the table.

What is the rule for finding the amount of an annuity in arrears What is the rule for finding the present worth of an annuity?

EXAMPLES.

8. What is the present worth of an annuity of $100, to continue 10 years, at 6 per cent. ?

9. What sum must be invested to yield an annual income of $500, at 6 per cent., for 12 years ?

10. What sum must be invested to yield an annual income of $2500, at 5 per cent., for 30 years ?

329. To find the present worth of an annuity that is perpetual, or is to continue forever, —

RULE. *Divide the annuity by the interest of $1 for 1 year.*

EXAMPLES.

11. What sum must be invested at 6 per cent., to yield an annual income of $500 ?

12. What sum must be invested at 5 per cent., to yield an annual income of $1200 ?

13. A farm rents for $450 a year. For what sum should it be sold, when money is worth 5 per cent. ?

14. A store rents for $1200 a year. For what sum should it be sold, when money is worth 6 per cent. ?

330. To find the present worth of an annuity in reversion, —

RULE. *Find the present worth of the annuity for the time before it commences, and also the present worth of the annuity from the present time till it terminates. The difference between these two present worths will be the present worth required.*

EXAMPLES.

15. What is the present worth of an annuity of $400 in reversion, at 5 per cent., commencing in 10 years, and continuing 20 years afterwards ?

What is the rule for finding the present worth of a perpetual annuity ? What is the rule for finding the present worth of an annuity in reversion ?

16. What is the present worth of an annuity of \$500 at $5\frac{1}{2}$ per cent., to commence in 10 years, and to continue 15 years afterwards?

331. To find an annuity, the present worth being given, —

RULE. *Divide the present worth by the worth of an annuity of \$1 for the time and rate.*

EXAMPLES.

17. What annuity, at 5 per cent., will pay a debt of \$10,000 in 4 years?
18. What annuity, at 6 per cent., will pay a debt of \$15,000 in 12 years?

332. To find an annuity, the amount being given, —

RULE. *Divide the amount by the amount of an annuity of \$1 for the given time.*

EXAMPLES.

19. What sum must be invested annually to amount to \$5000 in 10 years, at 6 per cent.?
20. What sum must be invested annually to amount to \$10,000 in 12 years, at 5 per cent.?
21. What annual sum must be invested as a sinking fund to pay a debt of \$20,000 in 10 years, at 6 per cent.?
22. What annual investment will amount to \$8000 in 6 years, at $5\frac{1}{2}$ per cent.?
23. What sum must be invested annually to amount to \$80,000 in 12 years, at 7 per cent.?

What is the rule for finding an annuity, when the present worth is known? What is the rule for finding an annuity, when the amount is known?

EXCHANGE OF CURRENCIES.

SECTION XXXVI.

333. EXCHANGE OF CURRENCIES is changing the denomination of one currency to that of another of equivalent value.

334. Accounts were kept in pounds, shillings, pence, &c., previous to the adoption of federal money in 1786. It is customary with many merchants at the present time to estimate the value of their goods in the old currency of the state.

335. Different values are assigned to the pound, &c., in the different states, as in the following table: —

\$1=6 s., or £$\frac{3}{10}$, in New England, Tennessee, Kentucky, Virginia, Mississippi, Missouri, Illinois, and Indiana.

\$1=7 s. 6 d., or £$\frac{3}{8}$, in New Jersey, Pennsylvania, Delaware, and Maryland.

\$1=8 s., or £$\frac{2}{5}$, in New York, Ohio, Michigan, and North Carolina.

\$1=4 s. 8 d., or £$\frac{7}{30}$, in South Carolina and Georgia.

\$1=5 s., or £$\frac{1}{4}$, in Canada and Nova Scotia.

Louisiana, Florida, Arkansas, and Alabama have no currency but federal money.

336. The old par value of a pound sterling was \$4.444; but, by a recent act of Congress, the present par value is \$4.844, which is 9 per cent. above the old par value. This is now the government value, and is adopted by all the custom-houses. The coin representing a pound is a sovereign.

OBS. The real value of a pound or sovereign of full weight is sometimes estimated at \$4.86$\frac{2}{3}$.

What is exchange of currencies? How were accounts formerly kept? What is the value of a dollar in the currency of each of the states? What is the currency of Louisiana, &c.?

337. To reduce federal money to any state currency, —

RULE. *Multiply the given sum by the value of $1 in the required currency, expressed in the fraction of a pound. The product will be pounds. Decimals must be reduced to shillings, pence, and farthings.*

1. Reduce $240 to New England currency.

$$240 \times \tfrac{3}{10} = £72.$$

EXAMPLES.

2. Reduce $560 to New York currency.
3. Reduce $780 to Pennsylvania currency.
4. Reduce $840.60 to Georgia currency.
5. Reduce $1200 to Canada currency.
6. Reduce $1600 to Maryland currency.

338. To reduce any state currency to federal money, —

RULE. *First reduce the shillings, pence, and farthings to the decimal of a pound. Then divide the pounds and decimals annexed by the value of a dollar in the given currency expressed in the fraction of a pound. The quotient will be dollars, cents, and mills.*

7. Reduce £72 New England currency to federal money.

$$72 \times \tfrac{10}{3} = \$240.$$

EXAMPLES.

8. Reduce £240, 10 s. New York currency to federal money.
9. Reduce £300, 12 s. 6 d. Pennsylvania currency to federal money.

What is the rule for reducing federal money to any state currency? What is the rule for reducing any state currency to federal money?

10. Reduce £460, 9 s. 10 d. Georgia currency to federal money.

11. Reduce £645, 16 s. Canada currency to federal money.

12. Reduce £1200, 15 s. New England currency to federal money.

339. To reduce federal money to English or sterling money, —

Rule. *Divide the given sum by the value of £1, ($4.844,) and point off the quotient as in the division of decimals,* (Art. 120.) *The figures on the left of the decimal point will be pounds, those on the right will be decimals, which must be reduced to shillings, pence, &c., as in* Art. 150.

13. Reduce $400 to pounds, shillings, &c.

400÷4.844=82.576=£82, 11 s. 6 d. 0.96 qr.

EXAMPLES.

14. Reduce $640 to pounds, shillings, &c.
15. Reduce $848 to pounds, shillings, &c.
16. Reduce $100 to pounds, shillings, &c.
17. Reduce $500 to pounds, shillings, &c.
18. Reduce $1000 to pounds, shillings, &c.
19. Reduce $1500 to pounds, shillings, &c.

340. To reduce English or sterling money to federal money, —

Rule. *Reduce the shillings, pence, and farthings to the decimal of a pound. Then multiply the legal value of a pound, ($4.844,) by the given number of pounds, and point off the product, as in the multiplication of decimals.*

What is the rule for reducing federal money to sterling money? What is the rule for reducing sterling money to federal money.

20. Reduce £100, 10 s. sterling to federal money.

$$100.5\times 4.84\tfrac{4}{9}=\$486.866.$$

EXAMPLES.

21. Reduce £240, 12 s. 6 d. sterling to federal money.

22. Reduce £500, 9 s. 9 d. sterling to federal money.

23. Reduce £640, 10 s. 8 d. sterling to federal money.

24. Reduce £784, 6 s. 4 d. sterling to federal money.

25. Reduce £2000, 10 s. 9 d. sterling to federal money.

EXCHANGE.

SECTION XXXVII.

341. EXCHANGE is the process of receiving or paying money, in one place, for its equivalent in another, by *order*, *draft*, or *bill of exchange.*

342. A *bill of exchange* is a written order drawn on a person in a distant place, requesting him to pay a certain sum of money, at a specified time, to another person, or his order. It is a mercantile contract in which three persons are primarily engaged, viz., the person who draws the bill, who is called the *drawer;* the person on whom it is drawn, who is called the *drawee;* the person to whom the money is to be paid, who is called the *payee.* When the person on whom a bill is drawn has accepted it, or engaged to pay it, he is called the *acceptor.*

What is exchange? What is a bill of exchange? How many persons are usually engaged in a bill of exchange? What is the person called who draws the bill? What, the person on whom it is drawn? What, the person to whom the bill must be paid? What, the person who accepts the bill?

343. The payee, or the person to whom the bill is to be paid, may, after having endorsed the bill, sell it to another, to be paid to him, or his order, and this person may, having endorsed it, sell it to a third, &c.

Obs. Endorsements are often made in blank, thus leaving a bill negotiable without further endorsement from parties through whose hands it passes.

344. The process of exchange is as follows: A, in Boston, owes B, in London, and C, in London, owes D, in Boston. A, instead of remitting the money to B, pays the money to D for a bill of exchange, or order, on C, requesting him to pay the sum named to A, or his order. A endorses the order, and sends it to B. Thus, by one process, A pays B, in London, and C, in London, pays D, in Boston, without the loss and risk of transporting the money.

345. Bills of exchange are either *foreign* or *domestic* — *foreign*, when drawn by a person in one country upon one residing in another; *domestic*, when both the drawer and drawee reside in the same country.

346. Foreign bills are usually drawn in sets of three each, which are sent by different conveyances, each one containing the condition that it shall be paid only while the others are unpaid.

Obs. By an act of Congress, passed in 1846, the foreign coins and money of account are to be estimated as follows: —

The specie dollar of Norway at	$1.06
The specie dollar of Denmark at	1.05
The thaler of Prussia and of the northern states of Germany at	.69
The florin of the southern states of Germany at	.40
The florin of the Austrian Empire and of the city of Augsburg at	.49
The ducat of Naples at	.80
The ounce of Sicily at	2.40
The lira of the Lombardo-Venetian kingdom, and the lira of Tuscany, at	.16
The franc of France and Belgium, and the lira of Sardinia, at	.18
The pound of the British provinces, Nova Scotia, New Brunswick, Newfoundland, and Canada, at	4.00

State the process of exchange.

A TABLE,

Showing the value of sterling money, in federal currency, from 1 pound to a penny.

£1, or 20 s.,	=	$4.844	4 s.	=	$0.968
19	=	4.60	3	=	.725
18	=	4.359	2	=	.484
17	=	4.17	1 s., or 12 d.,	=	.242
16	=	3.875	11	=	.222
15	=	3.630	10	=	.202
14	=	3.390	9	=	.182
13	=	3.148	8	=	.161
12	=	2.906	7	=	.141
11	=	2.664	6	=	.121
10	=	2.422	5	=	.101
9	=	2.178	4	=	·081
8	=	1.937	3	=	.061
7	=	1.695	2	=	.04
6	=	1.452	1	=	.02
5	=	1.21			

The government value of a pound has been assumed as the basis of the preceding table, and very small fractions have not been considered.

A TABLE OF EXCHANGE,

Showing the value of £1 sterling from 1 to 10½ per cent. above the old par value, $4.444.

Old par value,	$4.444	7 per cent. above,	$4.756
1 per cent above	4.489	8 " " "	4.80
1¼ " " "	4.50	8½ " " "	4.822
2 " " "	4.533	8¾ " " "	4.833
3 " " "	4.578	9 " " new par value	4.844
3½ " " "	4.60	9¼ " " "	4.856
4 " " "	4.622	9½ " " "	4.867
5 " " "	4.667	9¾ " " "	4.878
5½ " " "	4.68	10 " " "	4.89
5¾ " " "	4.70	10¼ " " "	4.90
6 " " "	4.711	10½ " " "	4.91

347. Exchange on England is usually reckoned on the old par value of the pound, $4.444.

348. To find the value of a pound sterling at any per cent. above par, —

RULE. *Multiply the number of pounds by the value*

What is the rule for finding the value of a pound at any per cent. above par?

of 1 pound at the given per cent., as shown in the preceding table.

OBS. Shillings, pence, and farthings must be reduced to decimals.

1. What is the value of £50 sterling at 10 per cent. above par?

$$50 \times 4.89 = \$244.50.$$

EXAMPLES.

2. What is the value of £240, 10 s. at 9½ per cent. above par?
3. What is the value of £500 at 8½ per cent. above par?

349. To change federal money to sterling money at any proposed per cent. above par, —

RULE. *Divide the given sum by the value of 1 pound sterling at the proposed per cent. above par, and point off as in the division of decimals. The figures on the left of the point will be pounds. The decimals must be reduced to shillings, pence, &c.*

4. What is the value of $490 in sterling money, at 10¼ per cent. above par?

$$490 \div 4.90 = £100.$$

EXAMPLES.

5. What is the value of $465 in sterling money, at 8 per cent. above par?
6. What is the value of $750 in sterling money, at 5¾ per cent. above par?
7. What is the value of $1000 in sterling money, at 9 per cent. above par?
8. What is the value of $1400 in sterling money, at 9¼ per cent. above par?

What is the rule for changing federal money to sterling money at any proposed per cent. above par?

Form of a Foreign Bill of Exchange.

Exchange for 6590 francs.

BOSTON, September 20, 1850.

Three months after sight, pay this my first of exchange (the second and third of the same date and tenor unpaid) to the order of Messrs. Brown and Co., the sum of six thousand five hundred and ninety francs, for value received, and place the same to my account, as advised.

GRENVILLE SMITH.

To Messrs. Stevens & Williams,
Bankers, Paris.

Form of a Domestic Bill of Exchange.

$5000. NEW YORK, September 20, 1850.

Seven days after sight, pay to Darius Alden, or order, the sum of five thousand dollars, for value received, with or without further advice.

HERMAN BATTLES.

To Messrs. Gates & Sons,
Merchants, New Orleans.

EXAMPLES.

9. What will a bill cost on England for £500, at 8 per cent. advance on the old par value?

10. What will a bill cost on England for £1000, 10 s., at 9 per cent. advance on the old par value?

11. What will a bill cost on France for 1600 francs, at 3 per cent. advance, estimating a franc at 18 cents?

12. What will a bill cost on New Orleans for $7560, at $1\frac{1}{2}$ per cent. advance?

13. What will a bill cost on Hamburg for 5000 marcs banco, at 2 per cent. above par, estimating a marc at 32 cents?

ARBITRATION OF EXCHANGE.

SECTION XXXVIII.

350. Arbitration of Exchange is the process of finding the rate of exchange between two countries, through the medium of several countries.

351. The object of the arbitration of exchange is to ascertain whether it would be more advantageous to the merchant to remit directly to the place where the money is due, or indirectly, through other countries.

352. Rule. *Write the first term on the left, and its equal on the right, and continue to write all the terms in pairs, in the same manner. The odd term, or term of demand, must be written under the terms on the left, if the answer is to be of the same kind as the first term; but if not, it must be written under the terms on the right.*

If the odd term be on the right, divide the product of all the terms on the right by the product of the terms on the left; but if the odd term be on the left, divide the product of the terms on the left by the product of the terms on the right.

Obs. 1. The second term on the left must be of the same kind as the first on the right, and the third term on the left of the same kind as the second on the right, &c., and the last term of the same kind as the first.

Obs. 2. The questions in arbitration of exchange may also be solved by analysis.

1. If the exchange of London on Hamburg be 1 pound sterling equal to 13 marcs and 14 schillings, and that of Hamburg on Paris be 100 marcs equal to 186 francs and 4 cents, how many francs in Paris,

---

What is arbitration of exchange? What is the object of arbitration of exchange? What is the rule?

through Hamburg, is a pound sterling, in London, worth? (*See note in the key.*)

Marcs. Schillings.
£1= 13 14

marcs 100=186.04 francs. 14 schillings=$\frac{7}{8}$ marc.

$$\frac{186.04\times 13\frac{7}{8}}{100}=\frac{2581.305}{100}=\text{25 francs, 81 cents.}$$

EXAMPLES.

2. If the exchange of London on Paris be 25 francs, 87 cents, worth 1 pound sterling, and that of Paris on Hamburg be 186 francs, 45 cents, worth 100 marcs, what is the rate of exchange of London on Hamburg, through Paris, or what is 1 pound sterling worth in Hamburg?

3. If A buys a bill in London, drawn on Paris, at the rate of 25.87 francs per pound sterling, and if this bill be sold in Amsterdam at 120 francs for 56$\frac{3}{4}$ florins, and the proceeds be invested in a bill on Hamburg, at the rate of 36$\frac{1}{4}$ florins for 40 marcs, what is the rate of exchange between London and Hamburg, or what is 1 pound sterling of London worth in Hamburg?

4. If the exchange of London on Hamburg be 13$\frac{1}{2}$ marcs per pound sterling, that of Hamburg on Amsterdam 40 marcs for 36$\frac{1}{4}$ florins, and that of Amsterdam on Paris 56$\frac{3}{4}$ florins for 120 francs, what is the rate of exchange between London and Paris, or how many francs are equal to 1 pound sterling?

5. If a bill be bought in London, drawn on Hamburg, at the rate of 13$\frac{1}{2}$ marcs per pound sterling, and if this bill be sold in Paris at 189$\frac{1}{2}$ francs for 100 marcs, and the proceeds be invested in a bill on Amsterdam, at the rate of 209 francs for 100 florins, what is the rate of exchange between London and Amster-

dam, or what is a pound sterling of London worth in Amsterdam?

6. If the exchange of London on Paris is 28 francs per pound sterling, and that of the United States on Paris 18 cents per franc, what is the rate of exchange of the United States on London, through Paris?

PERMUTATION.

SECTION XXXIX.

353. PERMUTATION is the process of finding how many changes can be made in the order of any given number of things.

354. RULE. *Multiply together in succession all the terms of the natural series of numbers, from* 1 *to the given number, and the last product will be the required number.*

EXAMPLES.

1. How many changes can be made in the positions of 6 persons around a table?
2. In how many different positions can the nine digits be placed?
3. How many changes can be made in the order of the 10 letters of the alphabet — A, B, C, D, E, &c.?
4. How many changes can be made in seating a class of 16 scholars?
5. A party of eight persons met at a public house, and agreed to dine together so long as they could sit in different positions at dinner. How long did they remain, and how much did they pay for their dinners, at 50 cents for each dinner?

What is permutation? What is the rule?

MENSURATION OF SURFACES.

SECTION XL.

355. The area or surface of a figure is the number of square rods, feet, inches, &c., it contains.

356. A *square* is a figure of four equal sides, which are perpendicular to each other.

357. A *rectangle* is a figure of four sides, which are perpendicular to each other, but the opposite sides of which only are equal.

358. To find the area of a square or rectangle, —

Rule. *Multiply the length by the breadth.*

EXAMPLES.

1. What is the area of a rectangular field, 40 rods in length and 30 rods in breadth?
2. What is the area of a field 50 rods square?
3. What is the area of a rectangular field, 60.5 rods long and 20.5 rods broad?
4. What is the area of a field 27.5 chains square?
5. How many square feet in a board 16 feet long and 1 foot and 4 inches wide?

359. In measuring land, surveyors use chain measure, of which a unit is a chain of 4 rods. A chain is divided into 100 links, and each link is 7.92 inches.

What is the area or surface of a figure? What is a square? What is a rectangle? What is the rule for finding the area of a square or rectangle? What measure is used by surveyors in measuring land? How many rods is a chain? How many links? What is the length of a link?

360. Any figure of four sides, whose opposite sides are equal, is a *parallelogram.*

361. To find the area of any parallelogram, —

RULE. *Multiply the length by the shortest distance between the two sides.*

OBS. The shortest distance between the two sides is a perpendicular line drawn between them.

6. What is the area of a parallelogram, whose longest side is 54 feet, and the shortest distance between the other two 24 feet?

362. A *triangle* is a figure having three sides and three angles.

363. To find the area of a triangle, —

RULE. *Multiply half the base by the perpendicular, or half the perpendicular by the base.*

7. What is the area of a triangle, whose base is 40 feet and whose perpendicular is 60 feet?

$(40 \div 2) \times 60 = 1200$ ft., or $(60 \div 2) \times 40 = 1200$ ft.

EXAMPLES.

8. What is the area of a triangle, whose base is 75 rods and whose perpendicular is 96 rods?

9. What is the area of a triangle, whose base is 60 yards and whose perpendicular is 80 yards?

What is the rule for finding the area of a parallelogram? What is the rule for finding the area of a triangle?

364. To find the area of a triangle, when the length of its three sides is known, —

RULE. *From half the sum of the three sides subtract each side separately. Then multiply the half sum by each side in succession. The square root of the continued product will be the area.*

10. What is the area of a triangle, whose sides are respectively 30, 24, and 18 feet?

$$(30+24+18)\div 2=36$$
$$36-30=6$$
$$36-24=12$$
$$36-18=18$$
$$\surd(18\times 12\times 6\times 36)=\surd 46656=216.$$

EXAMPLES.

11. What is the area of a triangle, whose sides are respectively 26, 34, and 16 feet?

12. What is the area of a triangle, whose sides are 40, 60, and 24 feet?

13. How many rods in a triangular piece of land, whose sides are 60, 44, and 18 rods?

365. A *trapezoid* is a figure bounded by four sides, only two of whose sides are parallel.

366. To find the area of a trapezoid, —

RULE. *Multiply half of the sum of the parallel sides by the shortest distance between them, or by a perpendicular line between them.*

What is the rule for finding the area of a triangle, when the length of its sides is known? What is a trapezoid? What is the rule for finding the area of a trapezoid?

OBS. A trapezoid and any parallelogram may be divided into two angles by drawing a line from one corner to its opposite.

EXAMPLES.

14. What is the area of a trapezoid, whose parallel sides are 40 and 60 feet, and the perpendicular, or shortest distance between them, 16 feet?

15. What is the area of a trapezoid, whose parallel sides are 96 and 36 feet, and the perpendicular 6 feet?

16. What is the area of a board 24 feet long, measuring 12 inches wide at one end and 16 inches at the other?

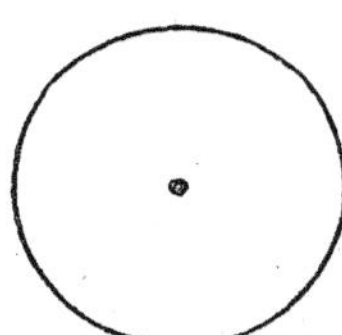

367. A *circle* is a figure bounded by a curve, every part of which is equally distant from a point called the *centre.*

368. The curve line bounding a circle is called the *circumference.* A straight line passing through the centre, and terminated by the circumference, is called the *diameter.* A straight line drawn from the centre, and terminated by the circumference, is called the *radius.*

369. The circumference of a circle is 3.1416 times the diameter.

370. To find the area of a circle, —

RULE. *Multiply half the diameter by half the circumference. Or multiply the square of the diameter by* .7854, *or the square of the circumference by* .0796.

17. What is the area of a circle, whose diameter is 40 rods?

$(40\times3.1416)\div2=62.8320.\ 62.8320\times20=1256.64.$

What is a circle? What is the circumference? What is the diameter? What is the radius? What is the rule for finding the area of a circle?

EXAMPLES.

18. What is the area of a circle, whose diameter is 64 rods?

19. What is the area of a circle, whose circumference is 130 rods?

20. What is the area of a circular pond, whose diameter is 20 rods?

21. What is the area of a circle, whose circumference is 48 feet?

371. To find the side of a square, equal in area to a circle whose diameter or circumference is given, —

RULE. *Multiply the diameter by* .8862, *or the circumference by* .2821.

EXAMPLES.

22. The diameter is 100 feet. What is the side of a square of equal area?

23. The circumference of a circle is 196 feet. What is the side of a square of equal area?

372. To find the side of the largest square inscribed in a circle of a given diameter or circumference, —

RULE. *Multiply the diameter by* .7071, *or the circumference by* .2251. *The product will be the side of the largest square.*

EXAMPLES.

24. What is the side of a square inscribed in a circle whose diameter is 16 feet?

25. What is the side of a square inscribed in a circle whose circumference is 156 feet?

What is the rule for finding the side of a square equal in area to a circle of a given diameter? What is the rule for finding the side of the largest square inscribed in a circle of a given diameter or circumference?

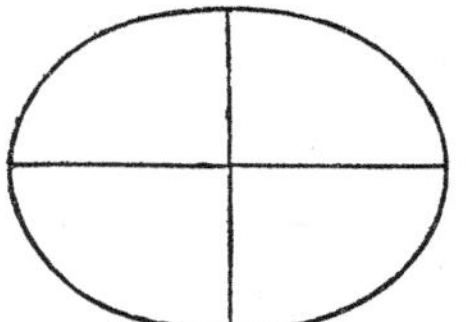

OBS. An *ellipse* may be formed by taking two points upon a plane surface and attaching to them a ring of thread and following it round with a pencil, keeping it extended in the form of a triangle.

373. To find the area of an ellipse, —

RULE. *Multiply the product of the longest and shortest diameter by* .7854.

EXAMPLES.

26. What is the surface of an elliptical pond, whose longest diameter is 100 feet, and shortest diameter 60 feet?

27. How many square feet in an elliptical table, the longest diameter of which is 4 feet, the shortest 2 feet, 9 inches?

MENSURATION OF SOLIDS.

SECTION XLI.

374. A SPHERE, or globe, is a solid, bounded by a curve surface, every point of which is equally distant from a point within it, called the centre.

375. The diameter, or axis, of a sphere is a straight line passing through the centre and both ends, terminating at the surface?

376. To find the surface of a sphere or globe, —

RULE. *Multiply the circumference by the diameter or axis.*

What is the rule for finding the area of an ellipse? What is a sphere? What is its diameter? What is the rule for finding its surface?

1. What is the area of a sphere whose diameter is 20 inches?

$$(20\times 3.1416)\times 20=1256.640=\text{the surface.}$$

EXAMPLES.

2. What is the surface of a sphere whose diameter is 30 inches?

3. What is the surface of a sphere whose circumference is 44 inches?

377. To find the solid content of a sphere,—

RULE. *Multiply the cube of the diameter by .5236.*

4. What is the solidity of a sphere whose diameter is 16 inches?

$$16^3\times .5236=2144.6656.$$

EXAMPLES.

5. What is the solidity of a sphere whose axis is 4 feet?

6. What is the solidity of the moon, supposing her to be a perfect sphere, and her axis 2180 miles?

378. To find the surface of a cylinder,—

RULE. *Multiply the circumference of the base by the height. The product will be the curve surface, to which add the areas of the ends.*

7. What is the surface of a cylinder, whose length is 20 feet, and the diameter of the base 30 inches.

$$3.1416\times 2\tfrac{1}{2}\times 20=157.080.$$

The area of the bases found by ART. 370.

8. What is the surface of a cylinder, whose length is 36 feet, and the circumference of its base 64 inches?

What is the rule for finding the solid content of a sphere? What is the rule for finding the surface of a cylinder?

379. To find the surface of a pyramid, —

RULE. *Find the area of the base, and the areas of the triangles which form the sides, separately. Their sum will be the whole surface.*

9. What is the surface of a triangular pyramid, each side of the base of which is 30 inches, and the perpendicular upon a side from the vertex 12 feet?

$(6\times2\frac{1}{2})\times3=45$ ft. = the surface of the sides.
The area of the base found by ART. 363.

10. What is the surface of a square pyramid, each side of the base of which is 40 inches, and the perpendicular upon a side from the vertex 16 feet?

380. To find the solidity of a pyramid, or cone, —

RULE. *Multiply the area of its base by one third of its height or length.*

EXAMPLES.

11. What is the solid content of a pyramid, whose height is 60 feet, and whose base is 20 feet square?

12. What is the solid content of a cone, whose height is 200 feet, and whose base is 50 feet in circumference?

381. To find the solidity of a prism, or cylinder, —

RULE. *Multiply the area of the base by its length or height.*

13. How many cubic feet in a cylinder, whose length is 10 feet, and the area of its base 4 feet?

$10\times4=40$ feet.

What is the rule for finding the surface of a pyramid? What is the rule for finding the solidity of a pyramid, or cone? What is the rule for finding the solidity of a prism, or cylinder?

EXAMPLES.

14. How many cubic feet in a stick of timber, 9 inches by 10 inches, and 50 feet in length?

15. How many cubic feet in a stick of timber, 10 inches by 14 inches, and 40 feet in length?

382. To find the surface of a frustum of a pyramid, or a cone, —

RULE. *Multiply half the sum of the circumference of both ends by the slant height, which will give the area of the curve, or convex, surface. The areas of both ends, added to this, will give the whole surface.*

16. What is the surface of a frustum of a square pyramid, the sides of the bases being 40 and 30 inches, and the slant height 12 feet?

$(40 + 30) \times 2 \times 12 = 1680$ in. = surface of slant height. $(40 \times 40) \div 12 = 133.3$ in. of one base. $(30 \times 30) \div 12 = 75$ in. of the other base. $(1680 + 133.3 + 75) \div 12 = 157.358$ square ft., whole surface.

EXAMPLES.

17. What is the whole surface of a frustum of a pyramid of five equal sides, the perpendicular height and the sides of the base at one end 36 inches, at the other 24 inches?

18. What is the whole surface of a frustum of a cone, the diameter of the larger base being 20 inches, of the smaller base 8 inches, and the slant height 12 feet?

383. To find the solidity of a frustum of a pyramid, or a cone, —

RULE. *Find the area of the two ends, and add the*

What is the rule for finding the surface of a frustum of a pyramid, or a cone? What is the rule for finding the solid contents of a frustrum of a pyramid, or a cone?

square root of their product to the sum of the areas of both ends. This, multiplied by one third of the perpendicular height, will give the solid content.

EXAMPLES.

19. What is the solid content of the frustum of a square pyramid, the sides of the bases being 15 and 6 feet, and the height 24 feet?

20. What is the solid content of the frustum of a triangular pyramid, the height of the frustum 14 feet, the sides of the greater base 21, 15, and 12, and those of the less base 14, 10, and 8 feet?

21. What is the solid content of the frustum of a cone, the diameters of the bases being 38 and 27 inches, and the height 11 feet?

22. What is the solid content of a mast of a ship, 60 feet high, and girths at one end 64 inches, and at the other 36 inches?

SIMILAR SURFACES.

SECTION XLII.

384. THE *areas of all circles are to each other as the square of their diameters, or as the square of their circumferences.* Thus, if the diameter of one circle be 3 feet, and the diameter of another be 4, the first circle is to the second as $3^2 : 4^2$, or as $9 : 16$; the first circle is $\frac{9}{16}$ of the second.

If the circumference of one circle be 6 feet, and that of another 10 feet, the first circle is to the second as $6^2 : 10^2$, or as $36 : 100$; the first circle is $\frac{36}{100} = \frac{9}{25}$ of the second.

What is the relation of different areas of circles?

EXAMPLES.

1. The diameter of a circle is 4 feet. What is the diameter of another, 6 times as large?

2. The diameter of a circular pond is 40 feet. What is the diameter of another, 5 times as large?

3. The circumference of a circle is 16 feet. What is the circumference of a circle 8 times as large?

4. The circumference of a pond is 150 feet. What is the circumference of a pond 6 times as large?

5. The circumference of a circular piece of land is 35 rods. What is the circumference of a piece $2\frac{1}{2}$ times as large?

385. *The areas of all similar surfaces are to each other as the squares of their like dimensions.*

386. *All solid bodies are to each other as the cube of their diameters or similar sides.*

6. If a ball 4 inches in diameter weigh 40 pounds, what will a ball of the same metal weigh, whose diameter is 6 inches?

$$4^3 : 6^3 :: 40 : 135.$$

EXAMPLES.

7. If a ball 6 inches in diameter weigh 135 pounds, what will a ball weigh whose diameter is 4 inches?

8. What must be the inside measure of a cubical bin containing 20 bushels? (268.8 cu. in. to the gall.)

9. What must be the inside measure of a cubical cistern, that will hold 50 hogsheads?

10. If a ball 10 inches in diameter weigh 240 pounds, what is the diameter of one of the same metal that weighs 45 pounds?

What is the relation of the different areas of all similar surfaces? What is the proportion of all solids to each other?

MENSURATION OF BOARDS AND TIMBER.

SECTION XLIII.

387. To find the area of a board, —

Rule. *Multiply the length by the mean width.*

Obs. One half of the sum of the width at each end is the mean width, when the board tapers regularly.

EXAMPLES.

1. What is the content of a board 9 feet, 6 inches long, and 14 inches wide at one end, and 12 inches wide at the other?

2. What is the content of a board 13 feet, 4 inches long, and 12 inches wide at one end, and 9 inches at the other?

3. What is the content of a board 10 feet, 10 inches long, and 11 inches wide?

388. To find the content of four-sided timber, —

Rule. *Multiply the mean breadth by the mean thickness. The product, multiplied by the length, will give the content.*

Obs. If the timber tapers regularly from one end to the other, one half the dimensions at the two ends may be taken for the mean dimensions. If the timber does not taper regularly, several dimensions may be taken at equal intervals, and their sum divided by the number of intervals, for the mean dimensions.

EXAMPLES.

4. What is the content of a stick of timber 24 feet, 6 inches long, the average breadth 14 inches, the average thickness 13 inches?

5. What is the content of a stick of timber 27 feet,

What is the rule for finding the area of a board? What is the rule for finding the content of four-sided timber?

4 inches long, the average breadth 18 inches, and the average thickness 15 inches?

389. To find the content of round timber, —

RULE. *Find one fourth of the mean girth, and square it, and multiply it by the length.*

OBS. An allowance must be made for the thickness of the bark No rough timber under 6 inches in diameter is measured in this way. This rule, though generally used, gives the content $\frac{1}{14}$ too small.

EXAMPLES.

6. What is the content of a tree 24 feet long, and its mean girth 8 feet?
7. What is the content of a tree 30 feet long, and its mean girth 5 feet, 8 inches?

GAUGING.

SECTION XLIV.

390. GAUGING is the process of finding the content of vessels of all kinds.

391. The dimensions, in gauging, are taken on the inside of the vessel.

392. RULE. *Add together the squares of the head, the bung, and twice the middle diameter. This sum, multiplied by the length, and also by* .000566, *gives the content in wine gallons; multiplied by* .0004642, *gives the content in beer gallons; by* .000487, *in dry gallons.*

OBS. 1. This is the most accurate method of finding the contents in casks.

OBS. 2. The middle diameter may be found by measuring the

What is the rule for finding the content of round timber? What is gauging? What is the rule for gauging?

circumference of the cask half the distance between the bung and the head, and dividing it by 3.1416, and then subtracting twice the thickness of the staves.

OBS. 3. A diagonal rod is sometimes used for finding the content of casks, but it does not give the contents so accurately as the preceding rule.

1. What is the content, in wine gallons, of a cask whose length is 40 inches, the bung diameter 34 inches, the head 27, and the middle diameter 32 inches?

$$(34^2+27^2+64^2)\times 40\times .0005667=135.577.$$

EXAMPLES.

2. What is the content of a cask, in wine gallons, measuring 48 inches in length, the head diameter 34, the middle 36, and bung 40 inches?

3. What is the content of a cask, in wine gallons, measuring 50 inches in length, the bung diameter 36 inches, the middle 34, and the head 32?

4. What is the content of a cask, in wine gallons, measuring 50 inches in length, the bung diameter 40 inches, the middle 36, and the head 34?

TONNAGE OF VESSELS.

SECTION XLV.

393. THE following rule is adopted by ship carpenters, who build vessels at a certain price per ton: —

RULE. *Multiply the length of the keel, breadth at the main beam, and depth of the hold together, and divide their continued product by* 95, *and the quotient will be the tonnage.*

In double-decked vessels, one half of the breadth is taken as the depth.

What is the carpenters' rule for finding the tonnage of vessels?

Government Rule. *If the vessel be double-decked, take the length from the fore part of the main stem to the after part of the main post, above the upper deck, the breadth at the broadest part above the main wales, half of which breadth shall be accounted the depth of such vessel, and deduct from the length $\frac{3}{5}$ of the breadth, and multiply the remainder by the breadth, and the product by the depth, and divide this last product by 95. The quotient will be the government tonnage.*

If the vessel be single-decked, multiply the length and breadth, as before taken, by the depth, taken from the under side of the deck to the ceiling in the hold.

EXAMPLES.

1. What is the tonnage of a single-decked vessel, by the carpenters' rule, whose keel is 75 feet, the breadth 24 feet, and the depth 8 feet?
2. What is the tonnage of a double-decked vessel, by the carpenters' rule, whose keel is 80 feet, and breadth 27 feet?
3. What is the tonnage of a single-decked vessel, by the government rule, which measures in length 84 feet, in breadth 32 feet, and in depth 12.5 feet?
4. What is the tonnage of a double-decked vessel, by the government rule, which measures at the keel 120 feet, and 36 feet breadth at the beam?

MECHANICAL POWERS.

SECTION XLVI.

394. There are six mechanical powers, viz., the *lever*, the *wheel and axle*, the *pulley*, the *inclined plane*, the *screw*, and the *wedge*.

What is the government rule for finding the tonnage of vessels? How many mechanical powers are there? What are they?

395. The *lever* is a bar, supposed to be inflexible, movable upon a *fulcrum.*

396. To find what weight can be raised by a given power, —

RULE. *The power is to the weight as the distance from the fulcrum to the weight is to the distance from the power to the fulcrum.*

EXAMPLES.

1. If the power be 150 pounds, the long arm 8 feet, and the short arm 2 feet, what weight can be raised?

OBS. The distance from the power to the fulcrum is called the *long arm*, the distance from the fulcrum to the weight, the *short arm.*

2. If the arms of a lever are 15 feet and 3 feet, and the weight 600 pounds, what is the power?

397. If the lever rest on two fulcrums, *the whole length is to the short arm as the whole weight is to the weight on the short arm;* or *the long arm is to the short arm as the weight supported by the short arm is to the weight supported by the long arm.*

EXAMPLES.

3. If A and B carry a weight of 300 pounds suspended upon a pole 9 feet long, and there is 5 feet between the weight and A, and 4 feet between the weight and B, how many pounds does each carry?

4. If A and B carry 150 pounds upon a lever 8 feet long, where must the weight be placed, that A may carry $\frac{2}{3}$ of it?

398. The principle of the wheel and axle is the same as that of the lever; the long arm of the lever

---

What is the lever? What is the rule for finding what weight can be raised by a given power? What is the principle of the wheel and axle?

corresponding to half the diameter of the wheel, and the short arm to half the diameter of the axle.

EXAMPLES.

5. If the diameter of a wheel be 4 feet, and that of the axle 6 inches, what weight will 160 pounds raise?

6. What weight will 300 pounds raise, if the diameter of a wheel be 10 feet, and that of the axle 2 feet?

399. In movable pulleys, *the power is to the weight as* 1 *is to twice the number of pulleys.*

EXAMPLES.

7. What weight can 300 pounds raise, with 3 movable pulleys?

8. What power can raise 2 tons, with 10 movable pulleys?

OBS. In the preceding examples no allowance is made for friction. In all practical applications, this allowance must be made.

400. On an inclined plane, *the length of the plane is to its perpendicular height as the weight is to the power.*

EXAMPLES.

9. An inclined plane is 30 feet long, its perpendicular height is 6 feet. What power will draw up a weight of 500 pounds?

10. What power will draw a train of cars weighing 50,000 pounds up an inclined plane, which rises 10 feet in 60 rods?

401. The principle of the screw is the same as that of an inclined plane. *The power is to the weight as the distance between two threads of the screw is to the circumference of a circle described by the power.*

What is the proportion of the power, in movable pulleys, to the weight? What is the rule for the inclined plane? What for the screw?

11. What weight must be applied to turn a screw whose threads are $\frac{1}{2}$ inch apart, with a lever 16 inches long, to raise 1000 pounds?

402. The *wedge* is a double inclined plane.

403. *The power applied to the head of the wedge is to the weight as one half the thickness of the head is to the length of its longest side.*

Obs. Great allowance must be made for friction, in all practical applications of the screw and the wedge.

GENERAL ANALYSIS.

SECTION XLVII.

404. It is very desirable that the pupil, after having learned the preceding rules, should be made familiar with applying the principles of analysis to all classes of questions. The pupil may be allowed to give a solution in his own way, but should be required to state each step in the process of the analysis.

1. If 24 yards of cloth cost \$120, what will 15 yards cost?

If 24 yards cost \$120, 1 yard will cost $\frac{1}{24}$ of $120 = \frac{120}{24}$. If 1 yard cost $\frac{120}{24}$, 15 yards will cost 15 times as much.

$$\frac{\overset{5}{\not{1}\not{2}\not{0}} \times 15}{\not{2}\not{4}} = \$75.$$

What is the rule for the wedge? What is it desirable that the pupil should be made familiar with?

2. If a family of 9 persons spend $120 in 8 months, how much will serve a family of 24, 16 months?

$$\frac{\overset{40}{\cancel{120}}\times\overset{2}{\cancel{16}}\times\overset{8}{\cancel{24}}}{\cancel{9}\times\cancel{8}}=\$640.$$

If 9 persons spend $120 in 8 months, 1 person would spend $\frac{1}{9}$ as much in 8 months, and $\frac{1}{8}$ as much in 1 month. 120 is therefore divided by 9 and 8. 24 persons would spend 24 times as much as 1 person in 1 month, and 16 times as much more in 16 months. Therefore 120 must be multiplied by 24 and 16. Cancelling the common factors, the answer will be $640.

EXAMPLES.

3. A person spent $\frac{1}{3}$, $\frac{1}{4}$, and $\frac{1}{5}$ of his money, and had $104 left. How much had he at first?

4. It is required to divide $620 among A, B, and C, so that A may have $\frac{1}{3}$, B $\frac{1}{5}$, and C $\frac{1}{15}$. What will each receive?

5. A says to B, If you will give me $50, I shall have as much as you. B says to A, Give me $44, and I shall have twice as much as you. How much had each at first?

6. A's age is double that of B, and B's is triple that of C. The sum of their ages is 140 years. What is the age of each?

7. A paid away $\frac{1}{4}$ and $\frac{1}{5}$ of his money, and had $66. How much had he at first?

8. Two persons, A and B, have the same income. A saves $\frac{1}{5}$ of his, annually; but B, by spending $50 per annum more than A, at the end of 4 years finds himself $100 in debt. What is their income?

9. Divide the number 36 into three such parts that

$\frac{1}{2}$ of the first, $\frac{1}{3}$ of the second, and $\frac{1}{4}$ of the third, may all be equal to each other.

10. If 240 men, in 5 days, of 11 hours each, can dig a trench 230 yards long, 3 yards wide, and 2 deep, in how many days, of 9 hours long, will 24 men dig a trench 410 yards long, 5 yards wide, and 3 deep?

11. If 6 pounds of tea be worth 8 pounds of coffee, and 12 pounds of coffee be worth 40 pounds of sugar, how many pounds of sugar may be obtained for 18 pounds of tea?

If 6 pounds of tea be worth 8 pounds of coffee, 18 pounds would be worth 3 times as much, 24. If 12 pounds of coffee be worth 40 pounds of sugar, 24 pounds would be worth 80 pounds of sugar. 80 pounds of sugar may be obtained for 18 pounds of tea.

12. If 18 bushels of corn are worth 3 barrels of flour, and 4 barrels of flour are worth 5 cwt. of sugar, and 6 cwt. of sugar are worth 3 cwt. of coffee, how much coffee can be purchased for 36 bushels of corn?

13. A, B, and C enter into partnership. A puts in $320, B $460, and C $500. They gain $200. What is each man's share of the gain?

14. A bankrupt owes $24650, and his whole property is worth only $16240. How much can he pay on $960.

15. A can do $\frac{2}{3}$ of a piece of work in 9 days, B can do $\frac{1}{2}$ of it in 4 days, and C can do $\frac{3}{4}$ of it in 12 days. In how many days can they do the whole, by working together?

16. A farmer had 84 sheep in two pastures. In one pasture there were twice as many as in the other. How many were there in each?

17. $\frac{4}{5}$ of a certain number exceeds $\frac{3}{8}$ of it by 17. What is the number?

18. A horse and saddle are worth $120. The horse is worth 5 times as much as the saddle. What is the price of each?

19. A horse and chaise are worth $350. The horse is worth $\frac{2}{5}$ as much as the chaise. What is the price of each?

20. There is a fish whose head is 8 inches long. His tail is as long as his head and half the length of his body; his body is as long as his head and tail both. What is the whole length of the fish?

21. A and B can cut 4 cords of wood in 3 days. B and C can cut 5 cords in 2 days. A and C can cut 7 cords in 4 days. How much can they cut in 1 day, working together?

22. A and B lay out equal sums of money in trade. A gains $126; B loses $87; and A's money is now double that of B's. What did each lay out?

23. What number is that, from which if 5 be subtracted, $\frac{2}{3}$ of the remainder will be 40?

24. What numbers are those, whose difference is 7, and sum 33?

25. Two travellers start at the same time from two places 150 miles apart, and travel towards each other. In what time will they meet, if one goes 8 miles a day, and the other 7?

26. A woman bought some apples at 3 for a cent, and as many more at 2 for a cent, and she sold them all again at 5 for 2 cents, and found she had lost 6 cents. How many of each did she buy?

27. What length must be cut off a board that is $7\frac{3}{4}$ inches broad, to contain a square foot, or as much as another piece 12 inches long and 12 inches broad?

28. Divide $1000 among A, B, and C, so as to give A $120 more and B $95 less than C.

PROBLEMS.

SECTION XLVIII.

405. The sum of two numbers, and the quotient resulting from one divided by the other, being given, to find the two numbers, —

Rule. *The sum of two numbers, divided by the quotient resulting from the greater divided by the less increased by* 1, *will give the less. The less, subtracted from the sum, will give the greater.*

1. The sum of two numbers is 36. If the greater be divided by the less, the quotient will be 5. What are the two numbers?

$36 \div \overline{5+1} = 6 =$ the less. $36 - 6 = 30 =$ the greater.

EXAMPLES.

2. Divide 144 into such parts, that if the greater be divided by the less, the quotient will be 24.

3. Divide 96 into such parts, that if the less be divided by the greater, the quotient will be $\frac{3}{5}$.

4. The sum of the ages of A and B is 60. A is three times as old as B. What are their ages?

5. Divide 156 into such parts, that if the greater be divided by the less, the quotient will be 26.

406. The difference of two numbers, and the quotient resulting from one divided by the other, being given, to find the two numbers, —

Rule. *The difference of two numbers, divided by*

What is the rule for finding two numbers, when the sum of the numbers and the quotient resulting from dividing one by the other are given? What is the rule for finding two numbers, when the difference of the numbers and the quotient resulting from dividing one by the other are given?

the quotient resulting from the greater divided by the less diminished by 1, *will give the less number. The less number, added to the difference, will give the greater.*

6. The difference between the ages of A and B is 12 years, but A is three times as old as B. What are their ages?

$12 \div \overline{3-1} = 6$, the less. $6 + 12 = 18$, the greater.

EXAMPLES.

7. The sum of the ages of A and B is 60 years, but B's age is $\frac{4}{5}$ of A's age. What are their ages?

8. A hare starts 60 rods before a greyhound, but the greyhound, pursuing the hare, runs 4 rods to the hare's 3. How many rods will the greyhound run before overtaking the hare?

9. The difference of two numbers is 92, and the smaller number is $\frac{1}{4}$ of the larger. What are the numbers?

407. The sum and product of two numbers being given, to find the two numbers, —

Rule. *Subtract the product of the two numbers from the square of half their sum. The square root of the remainder, added to half their sum, will be the larger number, which, subtracted from half their sum, will give the smaller.*

10. The sum of two numbers is 100, and their product is 2100. What are the numbers?

$(100 \div 2) \times 50 = 2500$. $\sqrt{(2500 - 2100)} = \sqrt{400} = 20$. $20 + 50 = 70$, the larger number. $50 - 20 = 30$, the smaller number.

What is the rule for finding two numbers, when their sum and product are given?

EXAMPLES.

11. The sum of two numbers is 9. Their product is 20. What are the numbers?

12. Divide 40 into two such parts that their product shall be 384.

13. Find two numbers whose sum shall be 30, and their product 224.

14. The sum of the ages of two brothers is 45 years, and the product of their ages is 500 years. What is the age of each?

408. The product and difference of two numbers being given, to find the two numbers, —

RULE. *Add the square of their difference to four times their product. The square root of the sum will be the sum of the two numbers. Half their sum, added to half their difference, will be the larger number.* (ART. 42.)

15. Find two numbers whose difference is 8, and their product 240.

$8^2 + \overline{240 \times 4} = 1024$. $\sqrt{1024} = 32$. $32 \div 2 = 16$. $16 + 4 = 20$, the larger number. $20 - 8 = 12$, the smaller.

EXAMPLES.

16. The difference of the ages of two brothers is 4 years. The product of their ages is 192 years. What is the age of each?

17. A says to B, The product of our ages is 2700 years, but their difference is only 4 years. How old was each?

18. The difference between the length and breadth of a rectangular field is 6 rods. The number of square

What is the rule for finding two numbers, when their product and difference are given?

rods in the field is 3240. What is the length and breadth of the field?

409. The sum of two numbers and the sum of their squares being given, to find the numbers, —

RULE. *Subtract the sum of the squares from the square of the sum. One half the remainder will be the product of the numbers. The sum and product being known, the numbers may be found by* ART. 407.

19. The sum of the squares of two numbers is 41. The sum of the numbers is 9. What are the numbers?

$\frac{81-41}{2} = 20$, the product. $\sqrt{(\frac{81}{4} - \frac{80}{4})} = \sqrt{\frac{1}{4}} = \frac{1}{2}$. $\frac{1}{2} + \frac{9}{2} = 5$, the larger number. $\frac{9}{2} - \frac{1}{2} = \frac{8}{2} = 4$, the smaller number.

EXAMPLES.

20. Divide 20 into two such parts that the sum of their squares may be 208.

21. Divide 36 into two such parts that the sum of their squares may be 656.

410. The sum of the squares of two numbers and the difference of the numbers being given, to find the two numbers, —

RULE. *From the sum of their squares subtract the square of their difference, and one half of the remainder will be the product of the numbers. The product and difference of the two numbers being known, the numbers may be found by* ART. 408.

22. The sum of the squares of two numbers is

What is the rule for finding two numbers, when the sum of the numbers and the sum of their squares are given? What is the rule for finding two numbers, when the sum of the squares and the difference of the numbers are given?

1424. The difference of the numbers is 12. What are the two numbers?

$(1424 - 12^2) \div 2 = 640$, the product. $(640 \times 4) + 144 = 2704$. $\sqrt{2704} = 52$. $52 \div 2 = 26$. $26 + 6 = 32$, the larger number. $32 - 12 = 20$, the smaller number.

EXAMPLES.

23. Find two numbers whose difference is 5, and the sum of their squares 1025.

24. There are two square fields, the sum of the contents of which is 596 rods. The difference in the length of the sides of each is 6 rods. What is the length of one of the sides of each field?

25. There were two laborers, each of whom dug a square piece of ground, of which the side was as many feet long as the laborer was years old. The difference of their ages was 12 years. The number of square feet dug by each was 1040 feet. What was the age of each laborer?

411. The sum of the squares of two numbers and the product of the numbers being given, to find the two numbers, —

RULE. *Add twice their product to the sum of their squares, and the square root of the sum will be the sum of the numbers. The sum and product being known, the numbers may be found by* ART. 407.

26. The sum of the squares of two numbers is 225. Their product is 108. What are the two numbers?

$\sqrt{(225 + \overline{2 \times 108})} = \sqrt{441} = 21$, their sum. $(21 \div 2)^2 = 10\tfrac{1}{2}^2 = 110\tfrac{1}{4}$. $110\tfrac{1}{4} - 108 = 2\tfrac{1}{4}$. $\sqrt{\tfrac{9}{4}} = \tfrac{3}{2}$. $\tfrac{3}{2} + \tfrac{21}{2} = 12$, the larger number. $21 - 12 = 9$, the smaller number.

What is the rule for finding two numbers, when the sum of the squares and the product of the numbers are given?

EXAMPLES.

27. Find two numbers whose product is 72, and the sum of their squares 145.

28. The product of two numbers is 200. The sum of their squares is 689. What are the numbers?

412. The sum of the cubes of two numbers and the sum of the numbers being given, to find the numbers, —

RULE. *Divide the sum of the cubes by the sum of the numbers, and subtract the quotient from the square of the sum.* $\frac{1}{3}$ *of the remainder will be the product of the numbers. The sum and product being known, the numbers may be found by* ART. 407.

29. The sum of the cubes of two numbers is 407. The sum of the numbers is 11. What are the numbers?

$407 \div 11 = 37$. $(121 - 37) \div 3 = 28$, the product. $\sqrt{(\frac{121}{4} - \frac{112}{4})} = \sqrt{\frac{9}{4}} = \frac{3}{2}$. $\frac{3}{2} + \frac{11}{2} = 7$, the larger number. $11 - 7 = 4$, the smaller number.

EXAMPLES.

30. The sum of the cubes of two numbers is 189. The sum of the numbers is 9. What are the numbers?

31. The sum of two numbers is 8, and the sum of their cubes is 152. What are the numbers?

32. The sum of two numbers is 21, and the sum of their cubes is 2331. What are the numbers?

413. The difference of the cubes of two numbers

What is the rule for finding two numbers, when the cubes and the sum of the numbers is given?

and the difference of the numbers being given, to find the numbers, —

RULE. *Divide the difference of the cubes by the difference of the numbers, and from the quotient subtract the square of the difference of the numbers.* $\frac{1}{3}$ *of the remainder will be the product of the numbers. The product and difference of two numbers being known, the numbers may be found by* ART. 408.

33. The difference of two numbers is 3, and the difference of their cubes is 117. What are the numbers?

$(117 \div 3) - 9 = 30$. $30 \div 3 = 10$, the product. $(10 \times 4) + 9 = 49$. $\sqrt{49} + 3 = 10$. $10 \div 2 = 5$, the larger number. $7 - 5 = 2$, the smaller number.

EXAMPLES.

34. The difference of the cubes of two numbers is 604, and the difference of the numbers is 4. What are the numbers?

35. The difference of the cubes of two numbers is 973, and the difference of the numbers is 7. What are the numbers?

36. The difference of the cubes of two numbers is 504, and the difference of the numbers is 6. What are the numbers?

37. The difference of the cubes of two numbers is 485, and the difference of the numbers is 5. What are the numbers?

38. There are two cubical stacks of hay. The larger contains 784 cubic feet more than the smaller. The difference of the sides of the stacks is 4 feet. What is the content of each stack?

What is the rule for finding two numbers, when the difference of the cubes and the difference of the numbers are given?

PRACTICAL QUESTIONS.

SECTION XLIX.

1. How many days are there from the fifteenth of May to the sixteenth of December, both days included?

2. The mariner's compass was invented in Europe in 1302. How long was that before the discovery of America by Columbus, in 1492?

3. There was an army composed of 104 battalions, each consisting of 600 men. What was the number of men in the whole army?

4. The salary of the President of the United States is $25,000 a year. What is that a day, there being 365 days in a year?

5. If one ounce of gold be estimated at $\$18\frac{2}{3}$, how many pounds of gold would there be in one million of dollars?

6. If one pound of silver be estimated at $\$15\frac{1}{2}$, what would be the weight of five millions of dollars in silver?

7. There are two numbers, the smallest of which is 80. Their difference is 28. What is the greater number, and what is their sum?

8. What number, subtracted from the square of 46, will leave 12 times 32?

9. There are two numbers, the greater of which is 73 times 109, and their difference 17 times 28. What is their sum and product?

10. What number is that, from which if $\frac{3}{8}$ of $\frac{4}{5}$ be subtracted, the remainder will be $\frac{10}{11}$?

11. What number is that, which being divided by $3\frac{4}{5}$, and multiplied by $4\frac{7}{8}$, the product will be 2?

12. How many bushels of oats, at 45 cents per bushel, must be given for 90 bushels of corn, worth 80 cents per bushel?

13. How many bushels of potatoes, at 67 cents per bushel, must be given for 2 barrels of flour, at $6 per barrel, 4 cwt. of sugar, at $6½ per cwt., and 16 pounds of tea, at 45 cents per pound?

14. What is the interest of $560 from May second, 1848, to March sixth, 1850, at 6½ per cent.?

LOWELL, September 4, 1840.

15. For value received, I promise to pay to William Snow, or order, $800, on demand, with interest.

LEONARD BOIT.

On this note were the following endorsements: —

January 6, 1842, $104. July 10, 1843, $16. September 9, 1843, $75. December 11, 1846, $120.

What was due April 15, 1848?

16. A merchant bought sugar at $6½ per cwt., cash. At what must it be sold, cash, to gain 20 per cent.?

17. A merchant bought sugar at $6 per cwt., 4 months' credit. At what must it be sold, cash, to gain 20 per cent., calculating interest for the time of credit at 6 per cent.?

18. A merchant bought sugar at $6 per cwt., cash. At what must it be sold per cwt. to gain 20 per cent., allowing 6 months' credit?

19. The sum of the ages of A and B is 72 years, but B's age is only ⅘ of A's age. What is the age of each?

20. There are two numbers, which are in the proportion of 5 to 8. The sum of the numbers is 32. What are the numbers?

21. There are two numbers, whose difference is 30.

The smaller number is $\frac{2}{5}$ of the larger. What are the numbers?

22. The difference between the ages of A and B is 6 years. B's age is $\frac{7}{8}$ of A's age. What are the ages of each?

23. Divide 50 into two such parts that their product shall be 525.

24. There is a piece of land in the form of a rectangular parallelogram, containing 1176 feet. The sum of two of the adjacent sides is 70 feet. What is the length and breadth of the field?

25. There is a piece of land in the form of a parallelogram, containing 960 feet, one side of which is 16 feet longer than the other side. What is the length of each side?

26. A is 8 years older than B, and the product of their ages is 2100 years. What is the age of each?

27. The difference of two numbers is 4. The sum of their squares is 656. What are the two numbers?

28. There are two square fields which together contain 1525 square feet. The side of the largest field is 5 feet longer than the side of the other. What is the length of one side in each field?

29. A merchant bought sugar at $6 per cwt., at 4 months' credit. At what price must it be sold to gain 20 per cent., and allow 6 months' credit?

30. A merchant bought sugar at $6 per cwt., cash. At what price must it be sold per cwt., cash, to gain 20 per cent., and allow 6 per cent. discount?

31. A merchant sold goods amounting to $1200, by which he gained 25 per cent. and allowed 10 per cent. discount. What was the prime cost?

32. Three persons, A, B, and C, purchased a ship,

of which A paid for $\frac{2}{9}$, B for $\frac{3}{5}$, and C paid \$1600. What part of the ship had C, and what did A and B pay?

33. A and B engaged in trade with a capital of \$4000, of which A's share was to B's as 7 to 5. At the end of 15 months, their whole stock was increased to \$5680. What was each partner's share of the gain?

34. If a person lends \$10,000 at 6 per cent. compound interest, and allows the interest to accumulate in the hands of the creditor, excepting \$250 per annum, which he wants for his family, how much will the creditor owe him at the end of 12 years?

35. A ship takes a cargo of sugar from Havana to St. Petersburg at £4 10 s. per ton of 2240 lbs., 5 per cent. primage. Her cargo consists of 2480 boxes, averaging 480 lbs. each, and tare averaging 54 lbs. each. What is the amount of her freight?

36. A ship was chartered for a voyage from Boston to Havana, and thence to St. Petersburg. She is to have \$1000 for the run from Boston to Havana, and £4 10 s. per ton, with 5 per cent. primage, for a full cargo of sugar from Havana to St. Petersburg. She delivered at St. Petersburg 495 tons. Her port charges at Havana were \$815, at St. Petersburg \$925. The expense of sailing was \$500 a month, and she was four months on the voyage. The wear and tear was estimated at 3 per cent., and the insurance paid was $2\frac{3}{4}$ per cent. on a valuation of \$25,000. What did she make on the voyage, estimating the pound sterling at \$4.80?

37. If two ships start from the same place at the same time, and one sails due south 56 miles, and the other due east 42 miles, how far apart will they then be?

38. There is a cone 40 feet high. How many feet of its top must be cut off to leave one half of its contents?

39. There is a square pyramid 60 feet high. What part of its top must be cut off to leave one third of its contents?

40. There are two cannon balls. One weighs 54 pounds, and the other 8 pounds. The diameter of the less is 4 inches. What is the diameter of the greater?

41. If a young man saves from his salary $100 a year, and invests it annually for 20 years, what sum will he have saved at that time?

42. Divide $300 among three persons, A, B, and C, so that when A receives $4, B receives $5, and C receives $6. How much will each receive?

43. What must be the radius of a circle that contains one acre?

44. Divide 72 into two factors, the sum of whose squares shall be 145?

45. If A's age be multiplied by B's age, the product will be 1200 years. The sum of the squares of their ages is 2500 years. What is the age of each?

46. There are two square buildings paved with stones a foot square each. The side of one building exceeds that of the other by 10 feet, and both pavements contain 2500 stones. What is the length of one side of each building?

47. There are two square fields which together contain 244 square rods, but one field contains 44 square rods more than the other. What is the length of one side of each field?

48. The difference of two numbers is 9. The difference of their squares is 351. What are the numbers?

49. There is a field in the form of a rectangular parallelogram, containing 192 square rods. The distance from one corner to its opposite is 20 rods. What is the length and the breadth of the field?

50. There are two numbers whose sum is 16. The sum of their cubes is 1792. What are the numbers?

51. There are two numbers whose difference is 10. The difference of their cubes is 3250. What are the numbers?

52. A sheepfold was robbed three nights in succession. The first night, half the sheep were stolen, and half a sheep more; the second night, half the remainder were taken, and half a sheep more. The third night, half of what were left were taken, and half a sheep more, when the number was reduced to 20. How many were there at first?

53. A young hare starts 40 rods before a greyhound, and is not perceived by him till she has been running 40 seconds. The hare runs at the rate of 10 miles an hour, and the dog pursues her at the rate of 18 miles an hour. How long will the dog be in overtaking the hare, and how far will he run?

54. A man, meeting a boy driving a flock of geese, said: Good morning, my boy; where are you going with your fifty geese? He replied, Sir, I have more than 50; and if I had $\frac{1}{2}$ as many more, and $\frac{1}{3}$ and $\frac{1}{6}$ as many more, I should have 180. How many had he?

55. What number is that which, being multiplied by 21, the product increased by 31, and the sum divided by 5, shall give 128 for the quotient?

56. What number is that which, if from three times its square 24 be subtracted, one half the remainder shall be equal to 3738?

57. A master, with his apprentice, can perform a piece of work in 8 days, which the master alone can perform in 12 days. In what time can the apprentice do it?

58. Divide 36 into two such parts, that the square of the greater shall exceed the square of the less by 360.

59. A person, inquiring the time of the day, received for an answer that the time past noon was $\frac{2}{5}$ of the time till midnight. What was the time?

60. The mast of a ship is 80 feet long. The diameter of the base is 3 feet, that of the top 8 inches. What is its solidity?

61. If a family of 10 persons spend $800 in 9 months, how many dollars would maintain them 12 months, if 6 persons were added to the family?

62. At what time, between five and six o'clock, are the hour and minute hands of a clock exactly together?

63. If 12 oxen will eat $3\frac{1}{3}$ acres of grass in 4 weeks, and 21 oxen will eat 10 acres in 9 weeks, how many oxen will eat 24 acres in 18 weeks, the grass being supposed to grow at an uniform rate during the time?

64. Two trees on a horizontal plane are 120 feet distant from each other. One tree is 100 feet and the other 80 feet high. How far from the base of each tree must a person stand, that the distance from the top of each tree to the point where he stands may be the same as the distance of the tops of the trees from each other?

65. If a person spend $\frac{5}{8}$ of his income in board and lodging, $\frac{1}{6}$ in clothes, and save $60 a year, what is his income?

66. In an orchard $\frac{1}{3}$ of the trees bear apples, $\frac{1}{4}$ pears, $\frac{1}{5}$ peaches; 50 bear cherries, and 90 plums. How many trees are there in the orchard?

67. What is the present value of a deferred annuity of $1000, to commence after the expiration of 5 years, and then to continue for 20 years, including compound interest at 6 per cent.?

68. From a vessel containing 10 gallons of wine 1 gallon was drawn out, and a gallon of water poured into the vessel. A gallon of the mixture was then drawn out, and another gallon of water poured in. Now, the like process being repeated 10 times, it is required to find how much wine remained in the vessel, supposing the two fluids were thoroughly mixed each time.

69. If a steam engine pass over 4 feet in the first second, and 88 feet in the sixtieth second of its motion, how far will it travel in the first minute, supposing its motion to be increased each second by a constant quantity?

70. The sum of three numbers in continued proportion is 52, and the sum of the extremes is to the mean as 10 is to 3. What are the numbers?

71. An army having been drawn up in a square, there were 79 men over; but in attempting to increase each side of the square by one man, there were wanting 80 men to complete the square. What was the number of men?

72. A gentleman has a garden in the form of a rectangle, surrounded by a walk 7 feet wide. There are 15,000 square feet in the garden, and 3696 square feet in the walk. What is the length and breadth of the garden?

FORMS OF PROMISSORY NOTES, DRAFTS, &c.

PROMISSORY NOTES.

$500. BOSTON, September 20, 1850.

For value received, I promise James Senter to pay him, or his order, five hundred dollars on demand, with interest. THOMAS GIBBS.

$800. BOSTON, September 30, 1850.

For value received, I promise George Wilson to pay him, or his order, eight hundred dollars in sixty days, with interest.

$600. SALEM, OCTOBER 10, 185

For value received, I promise Joseph Welsh to pa him, or his order, six hundred dollars in three month from date.

$400. LOWELL, September 10, 1850.

For value received, I promise Samuel Clark to pay him, or bearer, four hundred dollars on demand, with interest.

$640. BOSTON, June 12, 1850.

For value received, I promise to pay to my own order six hundred and forty dollars in sixty days from date. JAMES LEONARD.

OBS. When notes are made payable to order, they are negotiable, or may be passed to others, after being endorsed. When made payable to bearer, they are to be paid to the person who holds them, without endorsement. When made payable to a certain individual, named in the note, they are to be paid only to him.

DRAFTS.

$100. BOSTON, September 30, 1850.

At sight, without grace, pay to the order of James Stetson one hundred dollars, value received, and oblige Your obedient servant,

JOHN BENT.

Messrs. W. J. Reynolds & Co.,
Boston.

$360. BOSTON, October 3, 1850.

Sixty days after date, pay to the order of Edward Keyes three hundred and sixty dollars, value received, and place the same to my account.

WILLIAM TUDOR.

Messrs. Copeland & Kidder,
Boston.

RECEIPTS.

$250. BOSTON, September 6, 1850.

Received of William Bassett two hundred and fifty dollars on account.

CHARLES BLACK.

$460. BOSTON, June 10, 1850.

Received of Francis Jackson four hundred and sixty dollars in full of all accounts.

MANTON RANDALL.

$375. BOSTON, May 12, 1850.

Received of John Leonard three hundred and seventy-five dollars in full of all demands.

LEVI TOWER.

OBS. A receipt *on account* is given for part payment of a debt. A receipt *in full of all accounts* is given when all accounts are settled, and cuts off all accounts. A receipt *in full of all demands* cuts off all claims whatever but negotiable notes, which may be obtained at any time.

ORDERS.

BOSTON, May 9, 1850.

Mr. JOHN HUNT,

Please to pay Edwin Adams, or his order, forty-five dollars, value received, and charge the same to the account of Yours, &c.,

BENJAMIN LEMIST.

BOSTON, June 6, 1850.

Messrs. CLARK & HUNTINGTON,

Gentlemen, — Please to pay Samuel Putnam, or bearer, seventy-five dollars in goods, and charge the same to the account of

CALEB WOOD.

A FORM OF ACCOUNT CURRENT.

B. Flanner, Esq., in Acc't Curr't with Copeland & Kidder.

1849.	Interest to June 30, 1850.	Time.	Dr.	Int.	Cr.	Int.
		m. d.				
Aug. 28.	For balance acc't rendered, due this day,	10 2	737.92	37.14		
	By Sale No. 20, due Feb. 26, 1850,	4 2			163.14	3.53
	" Sale No. 21, cash this day,	10 2			113.37	2.69
	" Sale No. 19, due Dec. 4, 1849.	6 26			1016.39	34.89
	For allowance on Pitch, (Sale No. 19,) in bad order,		1.50			
Dec. 8.	" our acceptance of your draft, due March 4, 1850,	3 26	1250.00	24.17		
1850.						
Feb. 9.	By Sale No. 22, cash this day,	4 21			530.39	12.46
	" Sale No. 23, due April 19, 1850,	2 11			633.46	7.49
" 18.	For our acceptance of your draft, due May 19, 1850,	1 11	600.00	4.10		
June 10.	" balance of interest,		4.35			4.35
	By new account, for balance due June 30, 1850,				137.02	
			2593.77	65.41	2593.77	65.41
	For balance acc't, due June 30,		137.02			

Boston, June 10, 1850. Errors excepted,
Copeland & Kidder.

www.ingramcontent.com/pod-product-compliance
Lightning Source LLC
La Vergne TN
LVHW010230110826
845151LV00004B/1246

* 9 7 8 1 4 2 5 5 2 5 2 5 5 *